Methods in Molecular Biology

Series Editor
John M. Walker
School of Life and Medical Sciences
University of Hertfordshire
Hatfield, Hertfordshire, AL10 9AB, UK

For further volumes:
http://www.springer.com/series/7651

Hepatic Stem Cells

Methods and Protocols

Edited by

Naoki Tanimizu

Sapporo Medical University, Sapporo, Hokkaido, Japan

Humana Press

Editor
Naoki Tanimizu
Sapporo Medical University
Sapporo, Hokkaido, Japan

ISSN 1064-3745 ISSN 1940-6029 (electronic)
Methods in Molecular Biology
ISBN 978-1-4939-8960-7 ISBN 978-1-4939-8961-4 (eBook)
https://doi.org/10.1007/978-1-4939-8961-4

Library of Congress Control Number: 2018962407

Cover Illustration: Biliary epithelial cells (BECs) labeled with tdTomato proliferate to expand the CK19+ biliary epithelial tissue structure upon injury, indicating that pre-existing BECs contribute to the "ductular reaction". (The image related to Chapter 7 has been provided by Dr. Kenji Kamimoto).

This Humana Press imprint is published by the registered company Springer Science+Business Media, LLC, part of Springer Nature.
The registered company address is: 233 Spring Street, New York, NY 10013, U.S.A.

Preface

The liver contains two types of epithelial cells, namely, hepatocytes and cholangiocytes (biliary epithelial cells). Therefore, liver stem/progenitor cells (LPCs) are defined as bipotential cells that can differentiate into hepatocytes and cholangiocytes. Hepatoblasts, fetal LPCs, split into hepatocytes and cholangiocytes in mid-gestation. Bipotential LPCs also have been isolated from neonatal liver. On the other hand, it is still controversial whether bipotential LPCs contribute to maintenance of cellular homeostasis in healthy adult liver. Recent works demonstrated that both hepatocytes and cholangiocytes are heterogeneous cell populations and may contain committed progenitors. This book first contains methods for isolation and expansion of bipotential LPCs from fetal and neonatal liver, for adult clonogenic cholangiocytes and for hepatocyte progenitors.

The liver has the strong regenerative capability: it has been generally considered that self-duplication of hepatocytes or cholangiocytes compensates the lost tissue after acute injuries, whereas LPCs are activated and supply new hepatocytes after chronic injuries. However, recent works demonstrated that dedifferentiation and lineage conversion of liver epithelial cells also contribute to liver regeneration. These results highlight that liver regeneration can be achieved by multiple modes of cellular responses. In the second part, this book contains experimental methods for identifying and characterizing progenitor cells involved in liver regeneration in vivo as well as those for investigating molecular mechanisms regulating progenitor cell-driven liver regeneration.

It is crucial to expand functional hepatocytes and cholangiocytes ex vivo for implementing regenerative medicine such as cell therapy and for establishing drug screening systems. Recently, novel culture methods have been established to supply hepatocytes and/or cholangiocytes from pluripotent stem cells as well as somatic terminally differentiated cells. In addition, it was demonstrated that in vivo conversion of fibrogenic stellate cells into hepatocytes could be useful to ameliorate liver fibrosis. In the third part, this book contains culture methods for generating hepatocytes and cholangiocytes from multiple cellular sources and a technique converting fibroblasts into hepatocytes in vivo.

Liver epithelial cells form three-dimensional (3D) tissue structures, which are indispensable for the liver to perform their physiological functions. Facing various types of injuries, 3D hepatic tissue structures are dynamically rearranged to avoid fatal liver damages. Thus, this book also contains experimental methods for the reconstitution of 3D tissue structures ex vivo as well as for the investigation of the dynamic structural changes induced after injuries and during regeneration.

Fibrosis is the major pathology of the liver. Another issue is liver cancers. It is important to establish a novel strategy to resolve liver fibrosis and to understand etiologies of hepatocellular (HCC) and cholangiocellular carcinomas (CC). Therefore, the last part of this book includes methods resolving hepatic fibrosis by bone marrow transplantation as well as two novel mouse models of HCCs and CCs.

In addition to experimental protocols, each author provides a great introduction summarizing the background of each topic. This book would help researchers understand the current concept about hepatic stem/progenitor cells and perform basic and translational works related to liver biology and pathophysiology.

Sapporo, Hokkaido, Japan *Naoki Tanimizu*

Contents

Contributors

BEATRIZ AGUILAR-BRAVO • *Institut d'Investigacions Biomèdiques August Pi i Sunyer (IDIBAPS), Barcelona, Spain*

KAZUYA ANZAI • *Department of Gastroenterology, Tokai University School of Medicine, Isehara, Kanagawa, Japan*

PAMELA S. CANADY • *Oregon Stem Cell Center, Oregon Health and Science University, Portland, OR, USA*

HIROMI CHIKADA • *Department of Molecular Life Sciences, Tokai University School of Medicine, Isehara, Kanagawa, Japan*

ZHEN DAI • *Department of Gastroenterology, Hepatology and Endocrinology, Hannover Medical School, Hannover, Germany*

NICOLA DE PRISCO • *Cancer Biology Program, Fox Chase Cancer Center, Philadelphia, PA, USA*

CRAIG DORRELL • *Oregon Stem Cell Center, Oregon Health and Science University, Portland, OR, USA*

JOAN FONT-BURGADA • *Cancer Biology Program, Fox Chase Cancer Center, Philadelphia, PA, USA*

PENGYU HUANG • *School of Life Science and Technology, Shanghai Tech University, Shanghai, China*

LIJIAN HUI • *School of Life Science and Technology, Shanghai Tech University, Shanghai, China; State Key Laboratory of Cell Biology, CAS Center for Excellence in Molecular Cell Science, Shanghai Institute of Biochemistry and Cell Biology, Chinese Academy of Sciences, University of Chinese Academy of Sciences, Shanghai, China*

NORIHISA ICHINOHE • *Department of Tissue Development and Regeneration, Research Institute for Frontier Medicine, Sapporo Medical University School of Medicine, Sapporo, Japan*

MASAYUKI ISHII • *Department of Tissue Development and Regeneration, Research Institute for Frontier Medicine, Sapporo Medical University School of Medicine, Sapporo, Japan; Department of Surgery, Surgical Oncology and Science, Sapporo Medical University School of Medicine, Sapporo, Japan*

SEI KAKINUMA • *Department of Gastroenterology and Hepatology, Tokyo Medical and Dental University, Tokyo, Japan; Department of Liver Disease Control, Tokyo Medical and Dental University, Tokyo, Japan*

KENJI KAMIMOTO • *Department of Developmental Biology, Washington University School of Medicine in St. Louis, St. Louis, MO, USA*

AKIHIDE KAMIYA • *Department of Molecular Life Sciences, Tokai University School of Medicine, Isehara, Kanagawa, Japan; Center for Matrix Biology and Medicine, Graduate School of Medicine, Tokai University, Isehara, Kanagawa, Japan*

KOTA KANEKO • *Department of Pathology, School of Medicine, University of California, San Diego, CA, USA*

TAKESHI KATSUDA • *Division of Molecular and Cellular Medicine, National Cancer Center Research Institute, Tokyo, Japan*

TAKETOMO KIDO • *Laboratory of Cell Growth and Differentiation, Institute of Molecular and Cellular Biosciences, The University of Tokyo, Tokyo, Japan*

JUNICHI KINO • *Department of Tissue Development and Regeneration, Research Institute for Frontier Medicine, Sapporo Medical University School of Medicine, Sapporo, Japan; Tokushima Research Institute, Otsuka Pharmaceutical Co. Ltd., Tokushima, Japan*
SUNGJIN KO • *Department of Developmental Biology, McGowan Institute for Regenerative Medicine, Pittsburgh Liver Research Center, University of Pittsburgh, Pittsburgh, PA, USA*
KAZUHIKO KOIKE • *Department of Gastroenterology, The University of Tokyo, Tokyo, Japan*
NOBUHIKO KOJIMA • *Department of Life and Environmental System Science, Graduate School of Nanobioscience, Yokohama City University, Yokohama, Japan*
YUTA KOUI • *Laboratory of Cell Growth and Differentiation, Institute of Molecular and Cellular Biosciences, The University of Tokyo, Tokyo, Japan*
BIN LI • *Oregon Stem Cell Center, Oregon Health and Science University, Portland, OR, USA*
HIROTAKA MIHARA • *Department of Life and Environmental System Science, Graduate School of Nanobioscience, Yokohama City University, Yokohama, Japan*
TOSHIHIRO MITAKA • *Department of Tissue Development and Regeneration, Research Institute for Frontier Medicine, Sapporo Medical University School of Medicine, Sapporo, Japan*
HAYATO NAKAGAWA • *Department of Gastroenterology, The University of Tokyo, Tokyo, Japan*
YUJI NISHIKAWA • *Division of Tumor Pathology, Department of Pathology, Asahikawa Medical University, Asahikawa, Hokkaido, Japan*
TAKAHIRO OCHIYA • *Division of Molecular and Cellular Medicine, National Cancer Center Research Institute, Tokyo, Japan*
MICHAEL OTT • *Department of Gastroenterology, Hepatology and Endocrinology, Hannover Medical School, Hannover, Germany*
TERESA RUBIO-TOMÁS • *Institut d'Investigacions Biomèdiques August Pi i Sunyer (IDIBAPS), Barcelona, Spain*
YASUYUKI SAKAI • *Institute of Industrial Science, The University of Tokyo, Tokyo, Japan*
ISAO SAKAIDA • *Department of Gastroenterology and Hepatology, Yamaguchi University Graduate School of Medicine, Yamaguchi University, Ube, Yamaguchi, Japan*
PAU SANCHO-BRU • *Institut d'Investigacions Biomèdiques August Pi i Sunyer (IDIBAPS), Barcelona, Spain; Centro de Investigación Biomédica en Red de Enfermedades Hepáticas y Digestivas (CIBERehd), Barcelona, Spain*
AMAR DEEP SHARMA • *Department of Gastroenterology, Hepatology and Endocrinology, Hannover Medical School, Hannover, Germany*
XIZHONG SHEN • *Department of Gastroenterology, Zhongshan Hospital, Fudan University, Shanghai, China*
DONGHUN SHIN • *Department of Developmental Biology, McGowan Institute for Regenerative Medicine, Pittsburgh Liver Research Center, University of Pittsburgh, Pittsburgh, PA, USA*
GUANGQI SONG • *Department of Gastroenterology, Zhongshan Hospital, Fudan University, Shanghai, China*
ELEANOR STOUT • *Cancer Biology Program, Fox Chase Cancer Center, Philadelphia, PA, USA*
RYO SUDO • *Department of System Design Engineering, Keio University, Yokohama, Japan*
LULU SUN • *State Key Laboratory of Cell Biology, CAS Center for Excellence in Molecular Cell Science, Shanghai Institute of Biochemistry and Cell Biology, Chinese Academy of Sciences, University of Chinese Academy of Sciences, Shanghai, China*
NOBUMI SUZUKI • *Department of Gastroenterology, The University of Tokyo, Tokyo, Japan*

TARO TAKAMI • *Department of Gastroenterology and Hepatology, Yamaguchi University Graduate School of Medicine, Yamaguchi University, Ube, Yamaguchi, Japan*
KENJI TANI • *Department of Veterinary Surgery, Joint Faculty of Veterinary Medicine, Yamaguchi University, Yamaguchi, Yamaguchi, Japan*
NAOKI TANIMIZU • *Department of Tissue Development and Regeneration, Research Institute for Frontier Medicine, Sapporo Medical University School of Medicine, Sapporo, Hokkaido, Japan*
FUMIYA TAO • *Department of Life and Environmental System Science, Graduate School of Nanobioscience, Yokohama City University, Yokohama, Japan*
YASUHO TAURA • *Department of Veterinary Surgery, Joint Faculty of Veterinary Medicine, Yamaguchi University, Yamaguchi, Yamaguchi, Japan*
SHUJI TERAI • *Division of Gastroenterology and Hepatology, Graduate School of Medical and Dental Science, Niigata University, Niigata, Japan*
HSIN-CHIEH TSAY • *Department of Gastroenterology, Hepatology and Endocrinology, Hannover Medical School, Hannover, Germany*
ATSUNORI TSUCHIYA • *Division of Gastroenterology and Hepatology, Graduate School of Medical and Dental Science, Niigata University, Niigata, Japan*
KOTA TSURUYA • *Department of Gastroenterology, Tokai University School of Medicine, Isehara, Kanagawa, Japan*
LESLIE WAKEFIELD • *Oregon Stem Cell Center, Oregon Health and Science University, Portland, OR, USA*
BING XIN • *Division of Tumor Pathology, Department of Pathology, Asahikawa Medical University, Asahikawa, Hokkaido, Japan*
MASAHIRO YAMAMOTO • *Division of Tumor Pathology, Department of Pathology, Asahikawa Medical University, Asahikawa, Hokkaido, Japan*
QINGGONG YUAN • *Department of Gastroenterology, Hepatology and Endocrinology, Hannover Medical School, Hannover, Germany*
LUDI ZHANG • *State Key Laboratory of Cell Biology, CAS Center for Excellence in Molecular Cell Science, Shanghai Institute of Biochemistry and Cell Biology, Chinese Academy of Sciences, University of Chinese Academy of Sciences, Shanghai, China*

Part I

Isolation of Progenitor Cells

Chapter 1

Long-Term Culture of Mouse Fetal Hepatic Stem/Progenitor Cells

Atsunori Tsuchiya and Shuji Terai

Abstract

Mouse fetal liver includes abundant hepatic stem/progenitor cells (HSPCs). Easy expansion with passage of HSPCs is necessary to obtain steady data. However, it is often difficult to enrich only HSPCs, and HSPCs can die when usual trypsin is used for replating. Here, we introduce serum-free long-term culture with passage of HSPCs using fetal mouse liver without a cell sorter.

Key words Hepatic progenitor cell, Spheroid, Serum-free culture, Colony, Mouse

1 Introduction

Mouse fetal liver includes abundant hepatic stem/progenitor cells (HSPCs) surrounding hematopoietic cells, endothelial cells, mesenchymal cells, and other cells. Owing to the high expansion ability of HSPCs, it is very convenient to use these cells after expansion for in vitro and in vivo studies. There are two main problems in long-term culture of HSPCs. First, while HSPCs have a high expansion ability, other cells, such as hematopoietic cells, endothelial cells, and mesenchymal cells, also have high expansion ability. Sometimes when fetal liver cells are cultured without enrichment with serum, mesenchymal cells occupy most of the culture dish. Adopting a cell sorter is a very good approach to isolate only HSPCs by using the markers $CD45^{-}Ter119^{-}c\text{-}kit^{-}CD29^{+}CD49f^{+/low}$ [1, 2], epithelial cell adhesion antigen (EpCAM) [3], CD13 [4, 5], and delta-like 1 homolog (DLK1) [6, 7]. However, culture from single cells is not easy, and not all laboratories can apply this approach. Second, HSPCs can die when usual trypsin is used for replating. To overcome these problems and achieve long-term culture of HSPCs, we developed an approach involving spheroid culture followed by two-dimensional (2D) culture with replating using diluted trypsin [8] (Fig. 1).

Naoki Tanimizu (ed.), *Hepatic Stem Cells: Methods and Protocols*, Methods in Molecular Biology, vol. 1905,
https://doi.org/10.1007/978-1-4939-8961-4_1,

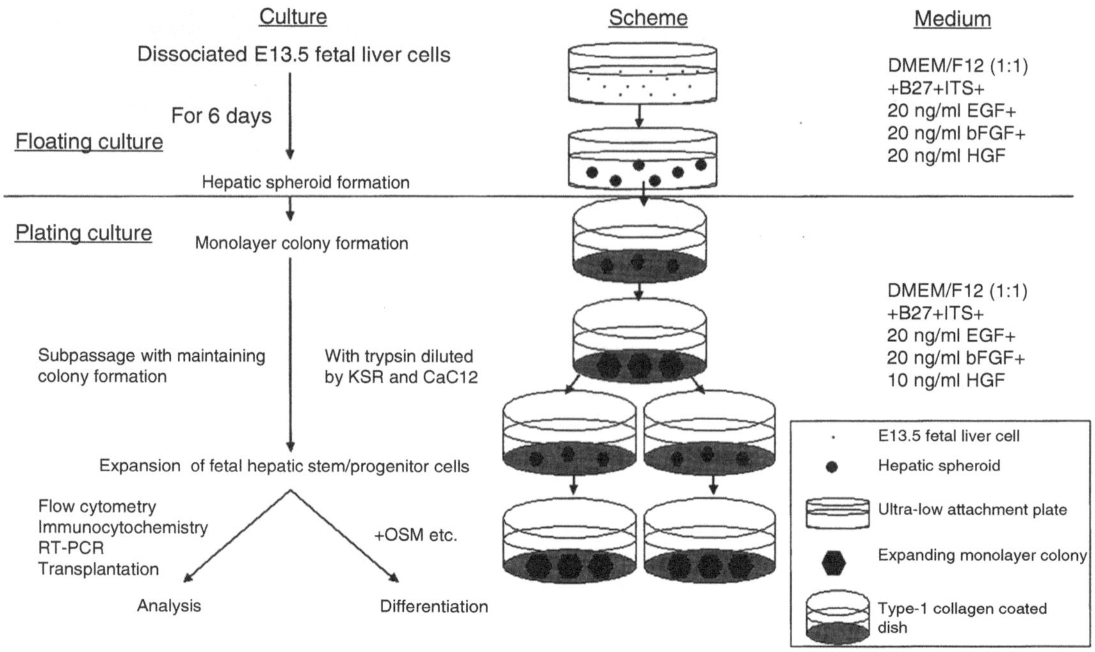

Fig. 1 Flow diagram presenting our culture system

2 Materials

2.1 Isolation of HSPCs

1. Mice: A pregnant C57BL/6 mouse at day 13.5 of gestation was used as a source of fetal liver. Other stains were not attempted for culture.
2. Tweezers.
3. Tissue culture dishes (100 mm, polystyrene).
4. Phosphate-buffered saline (PBS).
5. 1 mL pipette tips.

2.2 For Spheroid Culture (Floating Culture)

1. Medium: Dulbecco's Modified Eagle Medium/F12 (DMEM/F12).
2. Supplement: B27 supplement (50×), ITS mixture (100×), and HEPES (final concentration: 10 nmol/L) (*see* **Note 1**).
3. Antibiotics: Penicillin-streptomycin liquid (*see* **Note 1**).
4. Growth factors: Epidermal growth factor (EGF; final concentration 20 ng/mL), hepatocyte growth factor (HGF; final concentration 20 ng/mL), and basic fibroblast growth factor (bFGF; final concentration 20 ng/mL).
5. Dish: Six-well ultralow attachment plates (*see* **Note 2**).

2.3 For 2D Culture (Plating Culture)

1. Medium: Same as that in the spheroid culture.
2. Supplement: Same as that in the spheroid culture.
3. Antibiotics: Same as that in the spheroid culture.
4. Growth factors: Same as that in the spheroid culture. It is possible to reduce the concentration of each growth factor (20–10 ng/mL).
5. Dish: Six-well collagen type I-coated plates (*see* **Note 2**).

2.4 Diluted Trypsin

1. Trypsin (0.25%) 100 mL (final concentration 0.2%).
2. Knockout serum replacement (KSR) 25 mL.
3. $CaCl_2$ (final concentration 1 mmol/L) (*see* **Note 3**).
4. Diluted trypsin: Mix 1–3 to make diluted trypsin. Aliquot 15 mL to a Falcon tube, and freeze at −30 °C until use. Warm the Falcon tube at 37 °C before use.

3 Methods

3.1 Spheroid Culture

1. Before performing 2D culture, create spheroids using an ultralow attachment plate.
2. Kill a pregnant C57BL/6 mouse at day 13.5 of gestation using isoflurane, and perform cesarean delivery.
3. Remove and dissect fetal livers using tweezers in cold PBS.
4. Dissociate the dissected fetal livers using a 1 mL pipette tip, and collect the fragments with centrifugation at 500 rpm for 3 min (*see* **Note 4**).
5. Count and plate the collected cells (*see* **Note 5**) at a density of 5×10^5 cells/mL in the medium with growth factors mentioned in Subheading 2.2 on a 6-well ultralow attachment plate (2 mL/well).
6. Add 2 mL of medium and growth factors at days 2 and 4. Hepatic spheroids will form and increase in size for 6 days (Fig. 2) (*see* **Note 6**).

3.2 2D Culture

1. Collect the hepatic spheroids formed in Subheading 3.1 by centrifugation at 300 rpm for 2 min.
2. Plate the spheroids on a type I collagen-coated plate with the medium and growth factors mentioned in Subheading 2.2. Each spheroid will form a colony (Fig. 3a) in the 6-well collagen-coated plate (*see* **Note 7**). Hepatic progenitor cells will be highly enriched in this serum-free condition (*see* **Note 8**).
3. Change the medium and growth factors every 3 days.
4. After reaching approximately 80% of the confluency, collect cells according to the method mentioned below.

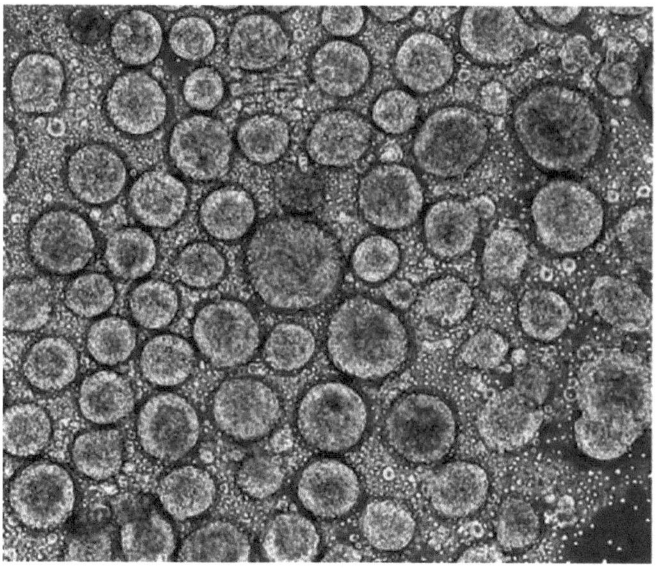

Fig. 2 Spheroids formed for 6 days by spheroid (floating) culture. The size of spheroids was approximately 50–250 μm, comprising approximately 20–150 cells

3.3 Replating

1. Remove the medium and wash with PBS.
2. Add 1 mL of diluted trypsin warmed at 37 °C to the 6-well collagen-coated plate, and warm the plate at 37 °C for 2–5 min.
3. Check the status of the colonies by microscopy every 2–3 min. If the colonies are partially removed from the plates while maintaining cell-cell contact, add medium and stop the enzymatic digestion/reaction (*see* **Note 9**).
4. Press a 2 mL or 5 mL pipette to the plates, and remove colonies with the strong hydraulic pressure of the pipette (Fig. 3b).
5. Collect the removed colonies with low-speed centrifugation (500 rpm for 3 min).
6. Divide the colonies into 2 or 3 aliquots, and replate on newly prepared collagen-coated plates (Fig. 3c). This procedure can be repeated to expand the mouse fetal HSPCs for more than ten passages. A scheme of the culture is shown in Fig. 1 (*see* **Note 10**).
7. Collected colonies can be preserved at −80 °C for several weeks or in liquid nitrogen for years by using STEM-CELLBANKER®. CELLBANKER® 2 can also be used (*see* **Note 11**).

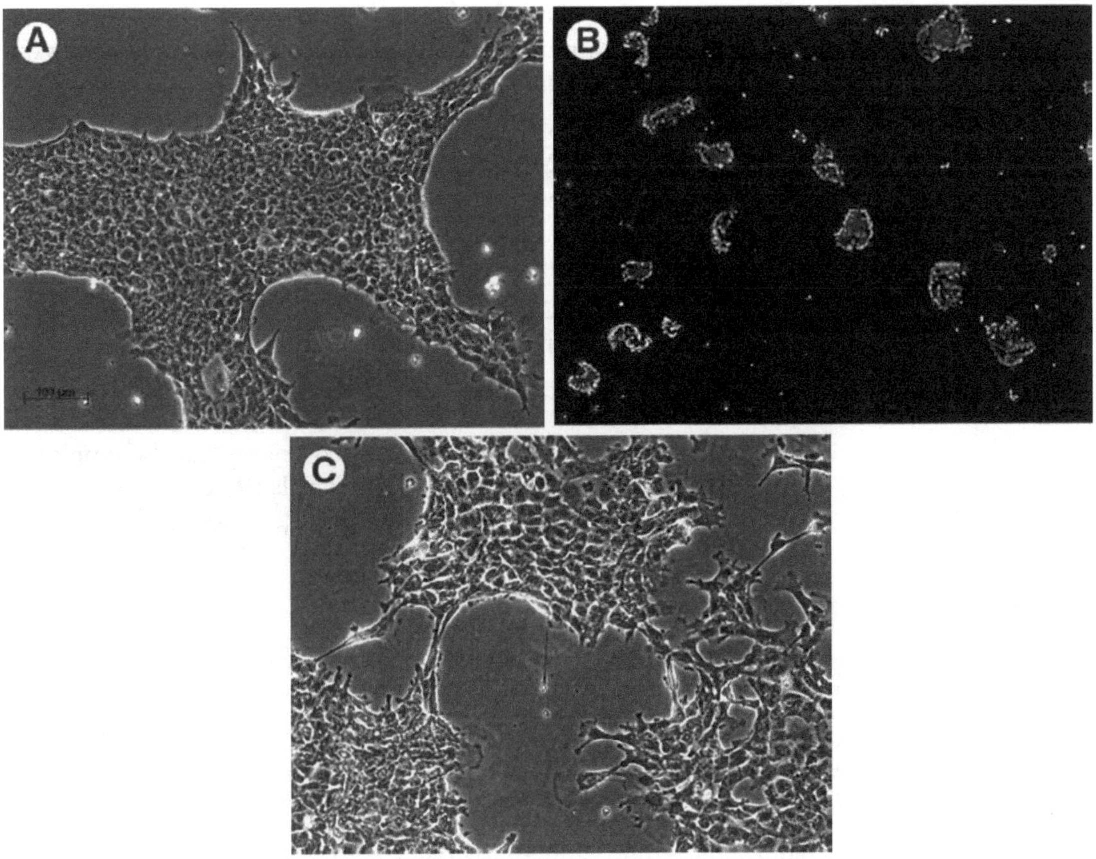

Fig. 3 Images of the 2D (plating) culture and replating. Colonies are formed from each spheroid (**a**). Collected colonies by diluted trypsin (**b**) are replated onto newly prepared collagen-coated plates (**c**). During passage, cell-cell contact was maintained

4 Notes

1. B27 supplement (50×), ITS mixture (100×), and penicillin-streptomycin liquid (100×) can be purchased from Thermo-Fisher Scientific.
2. Any type of plate may be appropriate if cells cannot attach to the plate. We purchased plates from Corning.
3. We made 100 mM $CaCl_2$ solution (100×) by dissolving $CaCl_2$ in distilled water and then filtering the solution. The solution was added to the mixture of trypsin and KSR.
4. Faster centrifugation might be better. However, 500 rpm for 3 min is sufficient to collect dissociated fetal liver cells.
5. During counting, red blood cells obviously distinguished by size were excluded.

6. We examined the efficacy to form spheroids through growth factor matching. Matching with EGF, HGF, and bFGF was the most appropriate to obtain many spheroids.
7. Spheroids from each 6-well plate were plated on each well of 6-well collagen type I-coated plates.
8. Hematopoietic cells and mesenchymal cells were included in the culture. However, most of the cells disappeared after 2–3 passages (confirmed by cell morphology and flow cytometry).
9. If colonies are difficult to remove from the plates, the time for trypsin treatment can be extended to 10–15 min. Maintaining cell-cell contact is the most important point for successful replating.
10. The replating method using diluted trypsin can be employed for adult mouse hepatic progenitor cells [9, 10].
11. The cell bankers can be purchased from ZENOAQ.

References

1. Suzuki A, Zheng Y, Kondo R et al (2000) Flow-cytometric separation and enrichment of hepatic progenitor cells in the developing mouse liver. Hepatology 32:1230–1239
2. Suzuki A, Zheng YW, Kaneko S et al (2002) Clonal identification and characterization of self-renewing pluripotent stem cells in the developing liver. J Cell Biol 156:173–184
3. Schmelzer E, Zhang L, Bruce A et al (2007) Human hepatic stem cells from fetal and postnatal donors. J Exp Med 204:1973–1987
4. Kamiya A, Kakinuma S, Yamazaki Y et al (2009) Enrichment and clonal culture of progenitor cells during mouse postnatal liver development in mice. Gastroenterology 137:1114–1126 1126 e1-14
5. Kakinuma S, Ohta H, Kamiya A et al (2009) Analyses of cell surface molecules on hepatic stem/progenitor cells in mouse fetal liver. J Hepatol 51:127–138
6. Tanimizu N, Nishikawa M, Saito H et al (2003) Isolation of hepatoblasts based on the expression of Dlk/Pref-1. J Cell Sci 116:1775–1786
7. Tanimizu N, Tsujimura T, Takahide K et al (2004) Expression of Dlk/Pref-1 defines a subpopulation in the oval cell compartment of rat liver. Gene Expr Patterns 5:209–218
8. Tsuchiya A, Heike T, Fujino H et al (2005) Long-term extensive expansion of mouse hepatic stem/progenitor cells in a novel serum-free culture system. Gastroenterology 128:2089–2104
9. Tsuchiya A, Heike T, Baba S et al (2007) Long-term culture of postnatal mouse hepatic stem/progenitor cells and their relative developmental hierarchy. Stem Cells 25:895–902
10. Lu WY, Bird TG, Boulter L et al (2015) Hepatic progenitor cells of biliary origin with liver repopulation capacity. Nat Cell Biol 17:971–983

Chapter 2

Isolation of Bipotential Liver Progenitor Cells from Neonatal Mouse Liver

Naoki Tanimizu

Abstract

Liver stem/progenitor cells (LPCs) are defined as bipotential cells differentiating into both hepatocytes and cholangiocytes. The Notch, TGFβ, and Hippo pathways have been implicated in lineage determination of LPCs during development and regeneration. However, the molecular mechanisms governing the lineage specification have not been fully elucidated, yet. Epithelial adhesion molecule (EpCAM) is a marker of cholangiocytes and of LPCs. We found that $EpCAM^+$ cells isolated from neonatal liver contain LPCs that clonally proliferate and are bipotential in vitro and in vivo. Furthermore, $EpCAM^+$ progenies keep the capacity of bidirectional differentiation even after long-term culture. These cells are useful to investigate the molecular mechanisms regulating lineage commitment and epithelial differentiation of LPCs.

Key words Tissue stem cell, Liver, Neonate, Bipotential, Hepatocyte, Cholangiocyte, Liver stem/progenitor cell, Fluorescence-activated cell sorting, Colony assay, Clonal culture

1 Introduction

Liver stem/progenitor cells (LPCs) are defined as bipotential cells differentiating into two types of liver epithelial cells, hepatocytes and cholangiocytes. LPCs have been isolated from normal and injured livers based on expression of surface antigens including EpCAM (epithelial adhesion molecule), CD13, and CD133 [1]. However, the genetic lineage tracing done in the past 5 years suggests that LPCs significantly contribute neither to cellular turnover nor regeneration [2, 3]. In contrast, recent results showed that cholangiocytes supply new hepatocytes when hepatocytes are almost completely depleted or the proliferative capability of hepatocytes is impaired [4, 5], suggesting that part of cholangiocytes may have plasticity to act as facultative LPCs. Although the heterogeneity of cholangiocytes has been demonstrated [6, 7], it is still difficult to label those cells in vivo and to prospectively isolate the sauce of facultative LPCs in normal and injured adult liver.

Naoki Tanimizu (ed.), *Hepatic Stem Cells: Methods and Protocols*, Methods in Molecular Biology, vol. 1905, https://doi.org/10.1007/978-1-4939-8961-4_2, © Springer Science+Business Media, LLC, part of Springer Nature 2019

We isolated $EpCAM^+$ cells from the liver between E17 and 12 W after birth and showed that colony-forming cells are enriched in this fraction [8]. We further found that neonatal $EpCAM^+$ cells differentiate to hepatocytes and cholangiocytes in vitro and in vivo, though they gradually lose such differentiation capability during the subsequent developmental process. It is also possible to establish bipotential liver progenitor cell lines from neonatal $EpCAM^+$ cells. This chapter includes the protocols for isolating $EpCAM^+$ cells from 1 W mouse liver, expanding them in vitro and inducing their differentiation to either hepatocyte or cholangiocytes.

2 Materials

2.1 Cell Isolation

1. Isoflurane.
2. Hanks' balanced salt solution (HBSS).
3. Pre-perfusion solution: 0.5 mM EGTA in HBSS.
4. Twenty-seven-gauge needle.
5. Ten milliliter syringe.
6. Collagenase: 1 mg/mL in HBSS.
7. 6 cm plastic dish.
8. Liver perfusion medium: Liver perfusion medium (Thermo Fisher Scientific, Cat. No. 17701038) or HBSS containing EGTA.
9. PBS (+/+):100 mg/mL $MgCl_2$ and 100 mg/mL $CaCl_2$ in phosphate buffer saline (PBS).
10. Liberase TM (Roche Diagnostics, Basel, Switzerland): It is the blendzyme of collagenase and thermolysin. Dissolve in PBS, make aliquots, and store at −80 °C. Dilute the stock solution 50 times in PBS (+/+) before use.
11. 70 μm cell strainer.
12. Hemolysis buffer: 10 mM Tris-HCl, 50 mM NH_4Cl. Pass through a 0.22 μm pore filter to sterilize.

2.2 Cell Sorting

1. Antibodies: FITC-conjugated anti-mouse CD329/EpCAM, APC-conjugated anti-mouse CD31, and APC-conjugated anti-CD45 and APC-Cy7-conjugated anti-mouse TER119.
2. Wash buffer: Add 2% FBS to PBS. Store at 4 °C.
3. Propidium iodide (PI) (1 mg/mL).

2.3 Cell Culture

1. Growth factor-reduced Matrigel® (MG) (BD Biosciences, Bedford, MA): Thaw the bottle of MG on ice, and make aliquots in 1.5 mL tubes. Freeze those tubes in liquid nitrogen, and store

them at −30 °C. Thaw MG in a tube on ice before use (*see* **Note 1**).

2. Collagen type IPC (Koken Co., Ltd., Tokyo, Japan).
3. Collagen reconstitution buffer: 2.2% $NaHCO_3$, 0.2 M HEPES, and 50 mM NaOH. Pass through a 0.22 μm pore filter to sterilize.
4. 10× Dulbecco's Modified Eagle's Medium (DMEM)/Ham's F-12 Nutrient Mixture (1:1) (DMEM/F-12) medium.
5. Laminin solution: Thaw laminin-111 in a glass bottle at 4 °C. Transfer it to a 1.5 mL tube, and store at 4 °C. Dilute it in PBS at 10 μg/mL. Add 1 mL diluted solution to 35 mm plate. Wait about 1 h and remove the solution before plating cells (*see* **Note 2**).
6. Gelatin solution: Autoclave PBS containing 0.1% gelatin. Add 300 μL gelatin solution to each well of 24-well plate. Aspirate gelatin solution and wash wells with PBS once.
7. Basal medium: Prepare DMEM/F-12 medium supplemented with 10% fetal bovine serum, 10 mM nicotinamide, and 10 μg/mL gentamicin.
8. Growth medium: Add 1× insulin/transferrin/selenium (ITS), 1×10^{-7} M dexamethasone (DEX), 10 ng/mL epidermal growth factor (EGF), and 10 ng/mL hepatocyte growth factor (HGF) to the basal medium.
9. Differentiation medium: Add 1× ITS, 1×10^{-7} M DEX, and 1% dimethyl sulfoxide (DMSO) to the basal medium. Add 10 ng/mL oncostatin M (OSM) or 5% MG before use.
10. Lysis solution for RNA extraction.
11. Cloning ring: Autoclave glass or stainless rings. Apply autoclaved Vaseline on the bottom of the ring before surrounding a colony.
12. Trypsin/EDTA: Dissolve 0.25% trypsin in PBS, make aliquots, and store at −30 °C. Dilute the stock solution 5 times with PBS and add with 1 mM EDTA before use.
13. Blocking solution: Dissolve Block Ace in PBS.
14. Bovine serum albumin (BSA).
15. PFS: Dissolve 0.7% fish gelatin, 0.07% saponin, and 0.02% sodium azide. Store at 4 °C.
16. Paraformaldehyde (PFA) solution: 4% PFA in PBS.
17. Permeabilization buffer: 0.2% Triton X-100 in PBS. Make 10% Triton X-100 and keep it at room temperature. Dilute the stock solution by 50 times with PBS.

3 Methods

3.1 Cell Isolation

3.1.1 Collagenase Perfusion

1. Anesthetize a mouse by isoflurane. Open the abdomen, and then cut the diagram and rib to expose the heart.
2. Attach a needle to a 10 mL syringe filled with pre-perfusion medium that is pre-warmed at 37 °C. Insert the needle into the left ventricle, and inject pre-perfusion medium. Cut the inferior vena cava after confirming the liver tissue becomes enlarged due to pressure (Fig. 1). Change the syringe to one filled with perfusion medium. Inject the perfusion solution slowly.
3. Isolate the liver and place it on a 6 cm plastic dish. Add 5 mL of HBSS. Hold the liver with a forceps, and scrape off hepatocytes from undigested liver tissue using another forceps. (Use undigested tissue for further enzymatic digestion. See the next paragraph.) Pass the cell suspension through a 70 μm cell strainer.
4. Centrifuge the cell suspension at 50 × *g* for 1 min. Collect the supernatant and centrifuge it at 350 × *g* for 4 min (*see* **Note 1**). Resuspend the pellet in 1 mL of basal medium.

3.1.2 Enzymatic Digestion of the Remaining Tissue After Collagenase Perfusion

1. Mince the undigested tissue after collagenase perfusion. Suspend tissue pieces in 3 mL of liver perfusion medium, and collect them in a 15 mL tube. Centrifuge at 150 × *g* for 3 min.
2. Resuspend the pellet in 1 mL of Liberase TM solution (*see* **Note 3**). Incubate the cell suspension at 37 °C for 15 min. Dissolve the tissue by pipetting (*see* **Note 4**). Add 1–2 mL of

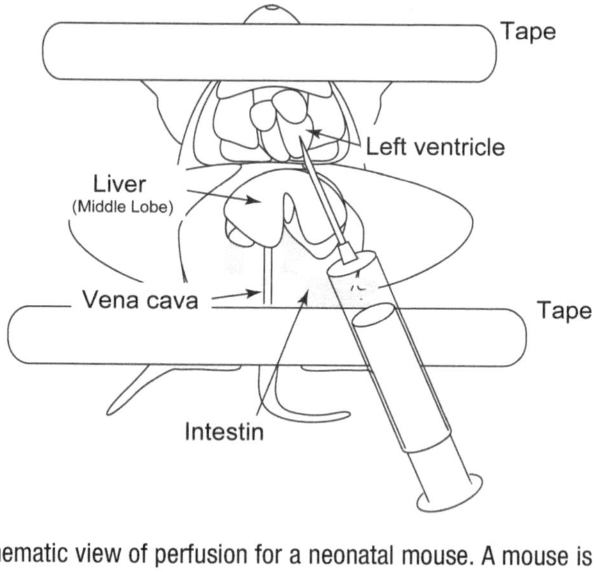

Fig. 1 Schematic view of perfusion for a neonatal mouse. A mouse is fixed on a board with tapes. After exposing the heart and identifying the left ventricle, a needle is attached to a 10 mL syringe filled with pre-perfusion medium

basal medium to stop enzymatic digestion. Pass the cell suspension through a 70 μm cell strainer.

3. Centrifugation at 50 × *g* for 1 min. Then, centrifuge the supernatant at 350 × *g* for 4 min to collect dissociated cells. Resuspend cells in 1 mL of basal medium.

3.1.3 Hemolysis

1. Combine the cell suspensions from Subheading 3.1.1, **step 4,** and Subheading 3.1.2, **step 3,** into a new 15 mL tube. Centrifuge at 350 × *g* for 4 min.
2. Resuspend pellet in 1 mL of hemolysis buffer. Incubate at 4 °C for 5 min. Then, add 1–2 mL of basal medium, and pass through a 70 μm cell strainer. Centrifuge at 350 × *g* for 4 min.
3. Resuspend the pellet in 200 μL of basal medium, and count the number of cells.

3.2 Cell Sorting

1. Centrifuge the cell suspension at 350 × *g* for 4 min. Resuspend cells in 100 μL of basal medium.
2. Add 1 μL of anti-CD16/CD32 antibody, and incubate at 4 °C for 20 min to avoid non-specific binding of antibodies by masking Fcγ receptors. Add ice-cold wash buffer and centrifuge at 350 × *g* for 4 min.
3. Resuspend cells in 100 μL of basal medium, and then add 1 μL each of FITC-conjugated anti-mouse EpCAM, APC-conjugated anti-CD31, APC-conjugated anti-CD45, and APC-Cy7-conjugated anti-TER119 (*see* **Note 5**).
4. For an isotype control, add FITC-conjugated rat IgG, APC-conjugated rat IgG, and APC-Cy7-conjugated rat IgG.
5. Incubate cells with antibodies at 4 °C for 20 min.
6. Add 2 mL of ice-cold wash buffer and centrifuge at 350 × *g* for 4 min. Resuspend cells in 300 μL of basal medium containing 1 μg/mL PI. Pass through 40 μm cell strainer before analysis on FACS Aria II (BD Biosciences).
7. Select $CD31^{-}CD45^{-}TER119^{-}$ cells in PI^{-} singlet cells by gating on FACS Aria II. Identify and collect $EpCAM^{+}$ cells in $CD31^{-}CD45^{-}$ fraction into a 5 mL tube (Fig. 2).
8. Transfer the collected cells into a 15 mL tube. Centrifuge at 350 × *g* for 8 min. Resuspend cells and add 7 mL of basal medium and centrifuge at 350 × *g* for 8 min. Resuspend cells in 1 mL of basal medium.

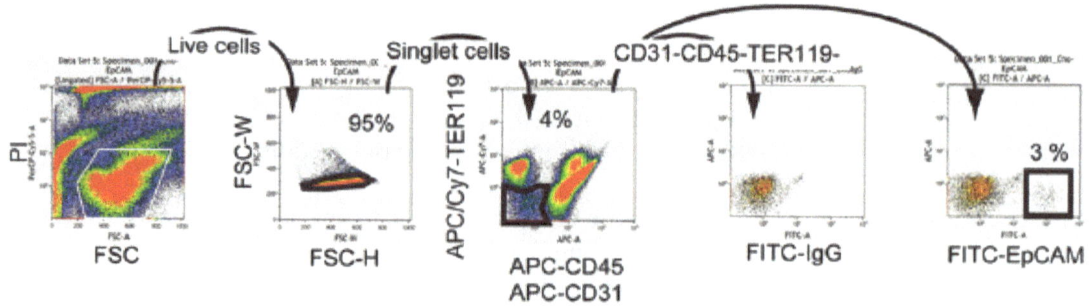

Fig. 2 Isolation of EpCAM$^+$ cells by FACS. After gating live singlet CD31$^-$CD45$^-$ cells, EpCAM$^+$ cells can be identified and isolated

3.3 Culture

3.3.1 Colony Assay

1. Resuspend cells in growth medium. Plate 5000 cells in 35 mm dish coated with laminin.
2. At day 7, fix cells in 4% PFA at 4 °C for 15 min. Wash 1 mL of PBS for 3 times. Permeabilize with 0.7 mL of 0.02% Triton X-100, and incubate at room temperature for 5 min. Wash cells with 1 mL of PBS for 2 times. Add 1 mL of Block Ace and incubate at room temperature for 30 min.
3. Replace blocking solution to 1 mL of PBS containing 1% BSA and 1 μL each of rabbit anti-mouse cytokeratin 19 (CK19) and anti-mouse albumin (ALB) antibody. Incubate at 4 °C for 4 h. Wash cells with PBS three times. Add PBS containing 1% BSA, 0.5 μL each of AlexaFluor488-conjugated anti-rabbit IgG and AlexaFluor555-conjugated anti-goat IgG and 1 μL each of Hoechst 33342. Incubate at 4 °C for 2 h. Wash cells with PBS three times before examination of the number of cells and expression of ALB and CK19 under a fluorescence microscope. A typical bipotential colony is shown in Fig. 3a.

3.3.2 Hepatocyte Differentiation

1. Plate 5000 EpCAM$^+$ cells in a well of 24-well plate coated with gelatin.
2. Replace basal medium to the differentiation medium containing OSM and incubate for 2 days. Replace the medium to one containing 5% MG. Keep the culture for additional 3 days.
3. To examine expression of hepatocyte marker genes, dissolve cells in a lysis solution and extract total RNA. Synthesize cDNA and perform PCR with gene-specific primers.
4. To examine expression of hepatocyte marker proteins, treat cells with Subheading 3.3.1, **step 2**. Incubate cells with rabbit anti-hepatocyte nuclear factor 4 α (HNF4α) and goat anti-ALB or with rabbit anti-CCAAT enhancer binding protein α (C/EBPα) and goat anti-carbamoyl phosphate synthetase I (CPSI) antibody. Typical images before and after the treatment of OSM and MG are shown in Fig. 3b.

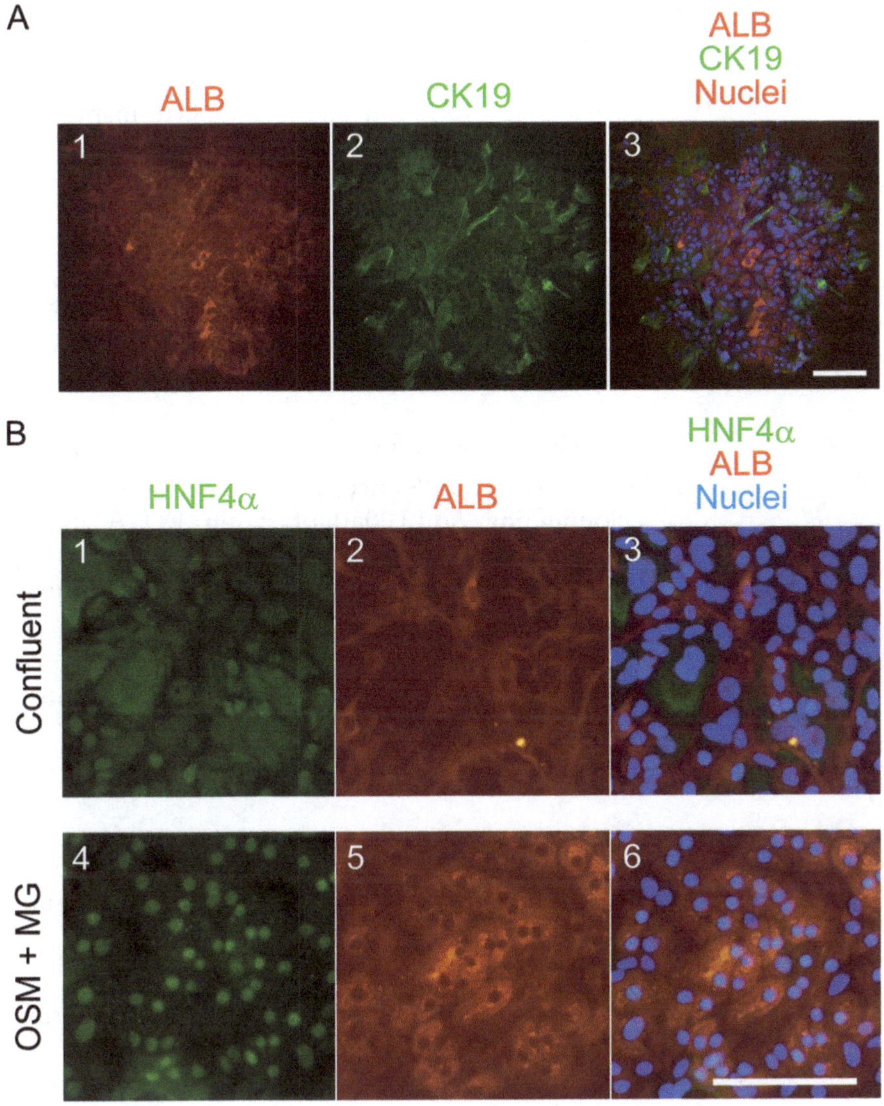

Fig. 3 Proliferation and hepatocyte differentiation of neonatal EpCAM^{+} cells. A EpCAM^{+} cell clonally proliferates and forms a bipotential colony containing ALB^{+} hepatocytes and CK19^{+} cholangiocytes. (**a**) Part of EpCAM^{+} cells expresses ALB and HNF4α before inducing hepatocyte differentiation, whereas they become HNF4α^{+}ALB^{+} hepatocytes in the presence of OSM and MG. (**b**) Bars represent 100 μm

3.3.3 Cholangiocyte Differentiation

1. Add PBS, collagen reconstitution buffer, 10× DF medium, and collagen IPC at 4:3:3:20 (v:v:v:v) ratio in a 15 mL tube on ice. Pipette up and down gently to mix thoroughly.
2. Mix collagen gel with MG at 1:1 ratio. Add 80 μL of the mixed gel to each well of 8-well cover glass chamber. Incubate it at 37 °C for 30 min.
3. Plate 1×10^4 cells in 150 μL of growth medium, and incubate for 5 min. Add 150 μL of growth medium containing 10% MG.

4. To examine expression of cholangiocyte marker genes, dissolve cells in a lysis solution, and extract total RNA. Synthesize cDNA and perform PCR with gene-specific primers.
   To examine expression of cholangiocyte marker proteins, fix cells in 4% PFA and permeabilize in 0.02% Triton X-100. Wash cells with PBS once, add PFS, and incubate at room temperature for 30 min. Incubate cells with PFS containing primary antibodies such as anti-CK19, anti-integrin beta4, anti-protein kinase C zeta, and anti-aquaporin 1 antibody followed by Alexa Fluor dye-conjugated antibodies and Hoechst 33342. Examine the localization of cholangiocyte markers under a confocal laser scanning microscope.

3.3.4 Isolation of Bipotential Clones

1. Keep clonal culture for about 4 weeks.
2. Wash cells with PBS once, and then surround each colony with a cloning ring. Add 100 μL of trypsin/EDTA, and incubate for 5–10 min at 37 °C. Pipette up and down for 3–4 times, and transfer the cell suspension into 1 mL of basal medium in 15 mL tube. Centrifuge at 350 × *g* for 3 min.
3. Resuspend cells in growth medium, and plate them onto a well of 24-well plate coated with laminin-111 (*see* **Note 6**).

4 Notes

1. It takes about 2 h to dissolve an aliquot of 500 μL MG on ice.
2. The diluted laminin solution can be reused up to three times to coat tissue culture dishes. We usually recover the laminin solution and store it in a 15 mL tube at 4 °C.
3. For scale-up, undigested tissues can be pooled at this step. The tissue pieces derived from 2 to 3 mice are collected in a 15 mL tube and then added with 2 mL of Liberase TM. For digesting tissue pieces from more than three mice, make another tube for Liberase digestion and then pooled cell suspensions into one tube after Liberase digestion.
4. Pipette up and down for 20–30 times to dissociate the most of cells from connective tissue.
5. Count the number of cells. Use 1 μL of each antibody to label cells less than 1×10^7 cells. Increase the quantity of antibodies according to the amount of cells.
6. The size of culture is gradually increased to 12-well plate and then 35 mm dish. EpCAM expression is examined by FACS to acquire EpCAM$^+$ clones. EpCAM$^+$ clones are replated every 4 days by plating 5×10^4 cells in 35 mm dish coated with laminin. Hepatocyte and cholangiocyte differentiation can be

induced using the same protocol as the primary $EpCAM^+$ cells, which are described in Subheadings 3.3.2 and 3.3.3.

Acknowledgments

I thank Dr. Toshihiro Mitaka and Dr. Norihisa Ichinohe for helpful discussion and Ms. Yumiko Tsukamoto and Ms. Minako Kuwano for technical assistance. This work is supported by the Ministry of Education, Culture, Sports, Science and Technology, Japan, Grants-in-Aid for Scientific Research (C) (25460271, 16 K08716), and Grants-in-Aid for Scientific Research on Innovative Areas "Stem Cell Aging and Disease" (17H05653).

References

1. Miyajima A, Tanaka M, Itoh T (2014) Stem Progenitor cell sin liver development, homeostasis, regeneration, and reprogramming. Cell Stem Cell 14:261–274
2. Espanol-Suner R, Carpentier R, Van HN et al (2012) Liver progenitor cells yield functional hepatocytes in response to chronic liver injury in mice. Gastroenterology 143:1564–1575 e1567
3. Malato Y, Naqvi S, Schurmann N et al (2011) Fate tracing of mature hepatocytes in mouse liver homeostasis and regeneration. J Clin Invest 121:4850–4860
4. Choi TY, Ninov N, Stainer DY et al (2014) Extensive conversion of hepatic biliary epithelial cells to hepatocytes after near total loss of hepatocytes in zebrafish. Gastroenterology 146:776–788
5. Raven A, Lu WY, Man TY et al (2017) Cholangiocytes act as facultative liver stem cells during impaired hepatocyte regeneration. Nature 547:350–354
6. Kamimoto K, Kaneko K, Kok CY et al (2016) Heterogeneity and stochastic growth regulation of biliary epithelial cells dictate dynamic epithelial tissue remodeling. Elife 5:e15034
7. Li B, DorrellC CPS et al (2017) Adult mouse liver contains two distinct populations of cholangiocytes. Stem Cell Reports 9:478–489
8. Tanimizu N, Kobayashi S, Ichinohe N et al (2014) Downregulation of miR122 by grainyhead-like 2 restricts the hepatocytic differentiation potential of adult liver progenitor cells. Development 141:4448–4456

Chapter 3

Identification and Isolation of Clonogenic Cholangiocyte in Mouse

Bin Li, Craig Dorrell, Pamela S. Canady, and Leslie Wakefield

Abstract

Cholangiocytes are proliferative and are one of the sources for liver progenitor cells. Clonogenic cholangiocytes are defined as cells capable of clonally proliferating and differentiating cholangiocytes both in vitro and in vivo. In this protocol, we describe the method for isolation of primary cholangiocytes from mouse. To study the heterogeneity of cholangiocytes, we used flow cytometry-based cell sorting to isolate different subsets of cholangiocytes. Organoid-forming efficiencies from sorted single cells are compared within different cholangiocyte populations to identify clonogenic cholangiocytes.

Key words Duct cells, ST14, FACS, Organoid

1 Introduction

Liver duct cells, so-called cholangiocytes, are highly proliferative in vitro and in vivo. In our previous report found in the normal and injured mouse liver, MIC1-1C3^{+} liver duct cells are proliferative: they formed colony in vitro, and gene expression showed MIC1-1C3^{+} cells that are expressing Sox9, a progenitor marker [1]. Moreover, MIC1-1C3^{+} colonies could engraft in the *Fah*$^{-/-}$ mouse, a transgenic mouse model for liver injury, and the transplanted liver cells could replace the damaged host liver [2], indicating the stem/progenitor capabilities of donor cells. Thus, MIC1-1C3 is used as an oval cell marker. Lgr5, a member of Wnt signal pathway, has been previously found as a liver stem cell marker. In contrast to MIC1-1C3, normal liver doesn't express Lgr5 [3], whereas, under chronic liver injury, liver duct cells in mouse become Lgr5^{+}. Those Lgr5^{+} cells are proliferative and form liver organoid in vitro. Moreover, transplantation of Lgr5^{+} organoids could engraft into the host liver, indicating the stem cell capability [4]. Further report demonstrated that MIC1-1C3^{+} duct cells in both normal liver and pancreas form organoids in vitro [5]. ST14 (suppressor of tumorigenicity 14 protein) is a marker for pancreatic cancer; however, not

Naoki Tanimizu (ed.), *Hepatic Stem Cells: Methods and Protocols*, Methods in Molecular Biology, vol. 1905, https://doi.org/10.1007/978-1-4939-8961-4_3, © Springer Science+Business Media, LLC, part of Springer Nature 2019

any research has studied ST14 in the liver. In our previous report, with the cell surface markers MIC1-1C3 [5] and ST14 [3], we demonstrated that normal mouse cholangiocytes are heterogeneous: ST14^{hi}MIC1-1C3^{+} cells are highly proliferative; they form single-cell-derived organoid more efficiently than ST14^{lo}MIC1-1C3^{+} cholangiocytes [3]. Here we will describe the method for isolating cholangiocytes via liver perfusion. ST14^{hi}MIC1-1C3^{+} cells in primary liver non-parenchymal cells (NPCs) are directly sorted into 96-well plate by fluorescence-activated cell sorting (FACS). In this format of colony-forming assay, clonogenic cholangiocytes form single-cell-derived organoids. In addition, organoids cultured for ~3 months are used for the differentiation assay examining their bidirectional differentiation potential.

2 Materials

2.1 Liver Perfusion

1. Solution 1: 10 mM HEPES and 0.5 mM EGTA in Earle's Balanced Salt Solution (EBSS) without Ca^{2+}/Mg^{2+}, pH 7.4 (~30 mL per mouse).
2. Solution 2: 10 mM HEPES in EBSS with Ca^{2+}/Mg^{2+}, pH 7.4 (~20 mL per mouse).
3. Solution 3: 0.1% collagenase II in solution 2 (~50 mL per mouse) *see* **Note 1**.
4. NPC digestion solution: 2.5 mg/mL collagenase IV and 0.1 mg/mL DNase I in solution 2 (~5 mL per mouse).
5. Wash medium: 10% FCS and 0.1 mg/mL DNase in DMEM.
6. 0.05% trypsin.
7. Phosphate-buffered saline (PBS).
8. 70% ethanol.
9. Cotton applicators (sterile).
10. Intravenous (IV) infusion set.
11. 24-gauge catheters.
12. Peristaltic pump.
13. Forceps, scissors (sterile).
14. Tape.
15. 100 μm cell strainer.
16. 60 mm petri dish.
17. Anesthetic solution.
18. Centrifuge, set at 10 °C.
19. Mouse: 8-week-old C57B/L6 male mouse. All animal experimentation was conducted in accordance with protocol IP00445 of the Institutional Review Committee at Oregon Health and Science University.

2.2 FACS

1. FACS buffer: 3% fetal bovine serum (FBS) in PBS.
2. 5% rat serum in PBS.
3. 5 mL polypropylene round-bottom tube.
4. 1 mg/mL propidium iodide (PI).
5. Antibodies: rat anti-mouse CD26, fluorescein isothiocyanate (FITC), CD45 Pe-cy7, CD31 Pe-cy7, CD11b Pe-cy7, MIC1-1C3 [1], 5% rat serum in PBS, rabbit anti-mouse ST14, secondary anti-rat phycoerythrin (PE), anti-rabbit allophycocyanin (APC).
6. Software for FACS data analysis: FlowJo software (FlowJo, LLC) or any other software that can handle FACS plots.

2.3 Organoid Cell Culture

1. 200 μL multichannel pipette and tips.
2. 96-well suspension culture plate (Genesee Scientific 25-104).
3. 24-well suspension culture plate (Genesee Scientific 25-102).
4. Matrigel (BD Biosciences) (*see* **Note 2**).
5. Organoid culture medium: advanced DMEM/F12 supplied with 100 ng/mL Noggin, 100 μg/mL Wnt3a, 50 ng/mL FGF10, 50 ng/mL EGF, 6 μM SB431542, 1.25 mM N-acetyl cysteine, 10 mM nicotinamide, 1× B27, 1× N-2 supplement, 1 μg/mL R-spondin1, 1× GlutaMAX, 10 mM HEPES, and penicillin/streptomycin/amphotericin.
6. TrypLE (Gibco).

3 Methods

3.1 Liver Perfusion

1. Sterilize pump tubing (at ~15 mL/min) with 70% ethanol and then wash with solution 1. Prewarm all solutions to 40 °C (*see* **Note 3**).
2. Attach IV infusion set and fill with solution 1 (*see* **Note 4**).
3. Anaesthetize the mouse.
4. Immobilize all four limbs with tape on a moisture-absorbing surface (diaper). Open the abdomen while avoiding opening the chest. Push intestines to the side with the sterile cotton applicator, thereby exposing the vessels. Make sure you identify the inferior *vena cava* and *portal vein* (under the liver) (Fig. 1a, b).
5. Carefully cannulate the *portal vein* (*see* Fig. 1c, d) (*see* **Note 5**). Fill the catheter with solution 1 by syringe, attach the IV tubing to the liquid-filled catheter (avoid introducing any air), and start the flow of the pump at 4 mL/min. The liver will fill up with solution and turn brown within a few seconds.

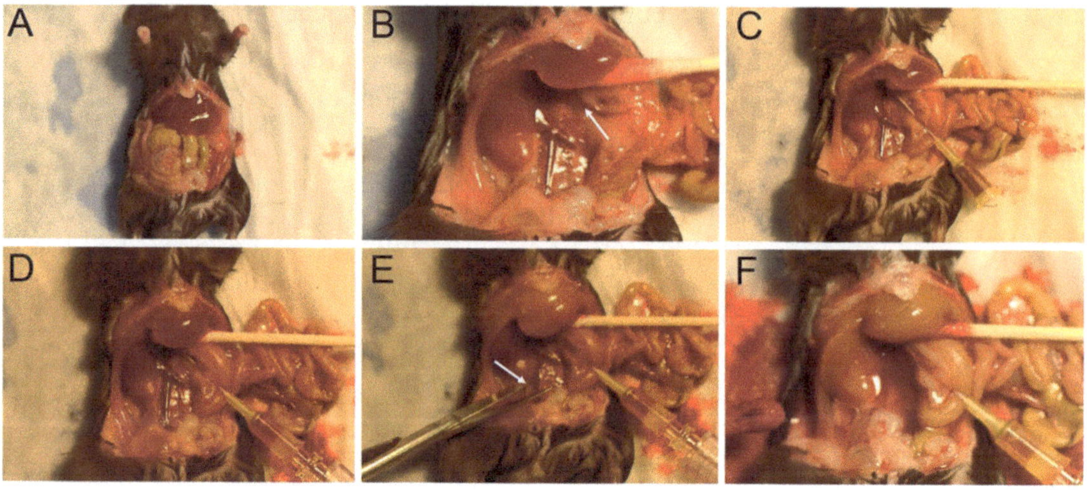

Fig. 1 Liver perfusion. (**a**) Anesthetize the mouse and open the abdomen. (**b**) Expose the *portal vein* with cotton applicator; white arrow shows the *portal vein*. (**c**) Cannulate the *portal vein*; blood will flow through the catheter. (**d**) Connect the tubing to the catheter. (**e**) White arrow shows cutting *vena cava*. (**f**) Perfused liver

Quickly cut the *vena cava* (*see* Fig.1e). Tape down the catheter and tubing securely. Start timer.

6. After about 15 mL of solution 1, run solution 2 for 1 min and finally solution 3 for 10 min (*see* **Note 6**).
7. To gauge progress, poke the liver with a cotton applicator after 7 min or so: under-digested liver remains elastic, whereas a completely digested liver will retain an indentation (Fig. 1f).
8. When the liver is digested, remove the catheter, excise the gallbladder, and then carefully remove the liver without injuring the capsule if possible.
9. Transfer the liver to a 60 mm petri dish containing solution 3, and proceed to the next perfusion (Fig. 2a). Disrupt the liver capsules with forceps, and wash out the hepatocytes by swirling the dish (Fig. 2b). Hepatocytes are collected and filtered through a 100 μm filter unit into a sterile 50 mL tube that contains 10 mL of wash medium (Fig. 2c).
10. For studies of non-parenchymal cells (NPC), tissue retained on the filter is collected and digested further.
11. Hepatocytes are pelleted by spin at 50 × *g* for 2 min at 4 °C (Fig. 2d). Supernatant is collected for NPC and the pellet resuspended in 50 mL wash media. Centrifuge at 300 × *g*, 5 min for NPC pellet; this is **NPC fraction 1**.
12. Transfer tissue fragments in **step 10** to a 60 mm dish in 5 mL of NPC *digestion solution*. Add a magnetic stirring bar, and stir at 37 °C for 10 min. *See* Fig. 2e.

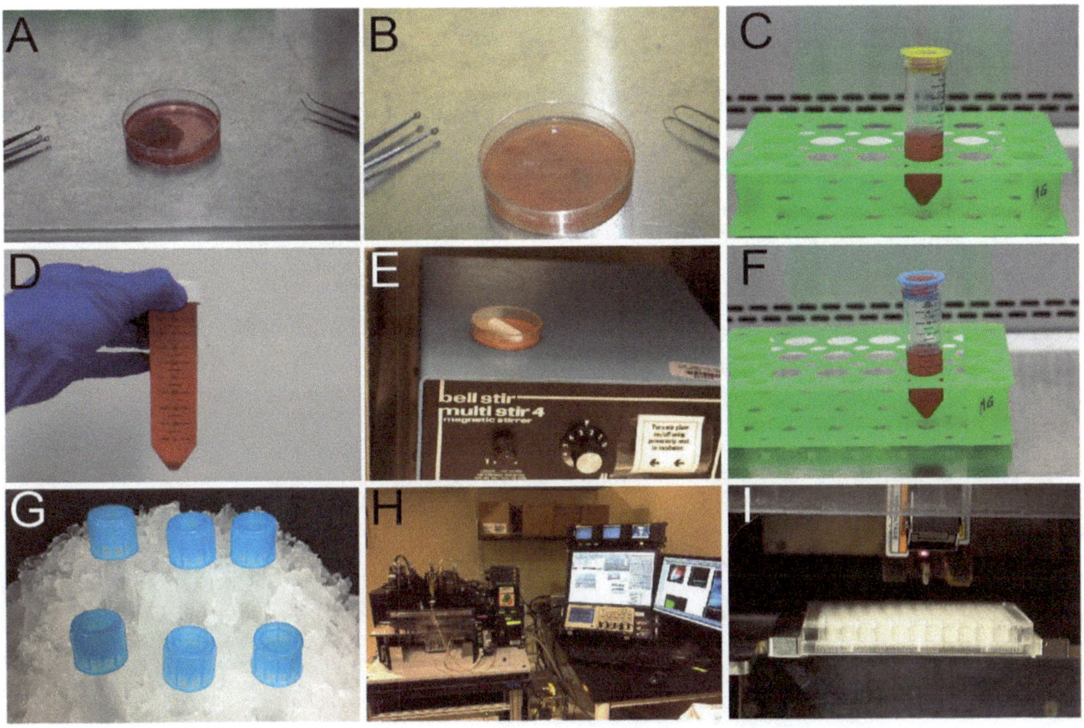

Fig. 2 Liver digestion and FACS sort. (**a**) Move the liver into a petri dish. (**b**) Break the liver capsule to dissociate the cells. (**c**) Filter the liver cells with cell strainer. (**d**) Pellet the hepatocytes. (**e**) Tissue leftover is under further digestion with NPC solution. (**f**) Filter the digested NPCs. (**g**) FACS tube. (**h**) FACS machine: BD Influx. (**i**) 96-well single cell sort

13. After 10 min, a P1000 is used to pipette solution up and down to assist the breakup of remaining tissue fragments.
14. Stirring is continued for an additional 5–10 min, and dissociated cells are collected by passage through a 40 μm strainer. This is **NPC fraction 2**. These should be spun at 300 × *g* and resuspended in cold wash media.
15. Remaining solid material is transferred to a new dish with 5 mL of 0.05% *trypsin* and incubated with stirring for 20 min.
16. A P1000 is used to pipette solution up and down to assist the breakup of remaining tissue fragments.
17. Newly dissociated cells are collected by passage through a 40 μm strainer. This is **NPC fraction 3**. These should be spun at 300 × *g* and resuspended in cold wash medium.
18. Collect all NPC fractions and pellet the cells.

3.2 FACS

1. Count and resuspend the cells with FACS buffer at a density of 1×10^7 cells/mL. Add 100 μL cells into each control tube (#1–7). The sample tube (#8) can have 500–1000 μL volume (*see* **Note 7**).

Table 1
Antibody labeling for FACS

Tube#	1	2	3	4	5	6	7	8
	PI	FITC	PE	PEcy7	APC	FM1 -APC	FM1 -PE	Sort
MIC1-1C3 (μL)			5			5		50
ST14 (μL)					1		1	10
FACS buffer (μL)			100	100	100	100	100	1000
20 min on ice, wash, pellet cells								
Anti-rat PE (μL)			1			1	1	2
Anti-rabbit APC (μL)					1	1	1	2
FACS buffer (μL)			200		200	200	200	400
15 min on ice, wash, pellet cells, blocking tubes #6 and #7 with 5% rat serum for 10 min								
FACS buffer (μL)		100		100		100	100	200
FITC CD26 (μL)		1				1	1	2
Pe-cy7 CD45/31/11b (μL)				1		1	1	2
Cell number	10^6	10^6	10^6	10^6	10^6	10^6	10^6	10^7

2. Add rat anti-mouse MIC1-1C3 antibody (1:20 dilution of each) to tubes # 3, 6, and 8. Add rabbit anti-mouse ST14 antibody (1:100 dilution) to tubes # 5 and 7.
3. Incubate 20 min on ice.
4. Wash (add 5 mL wash medium to tubes, spin 4 min at 300 × *g* to pellet, aspirate supernatant), and add 200 μL FACS medium.
5. Add goat anti-rat F(ab) PE (1:200 dilution) to tubes #3, 6, 7, and 8 and mouse anti-rabbit APC (1:200 dilution) to tubes #5, 6, 7, and 8.
6. Incubate 15 min on ice.
7. Wash (add 5 mL wash medium to each tube, spin 4 min at 300 × *g* to pellet, aspirate supernatant).
8. Block tubes #6, 7, and 8 with 100 μL and 200 μL 5% rat serum for 10 min on ice.
9. Add rat anti-mouse CD26 to tubes #2, 6, 7, and 8.
10. Add rat anti-mouse CD45/CD31/CD11b Apc-cy7 (1:100 dilution) to tubes #4, 6, 7, and 8. *See* Table 1 for antibody labeling. All steps are on ice.

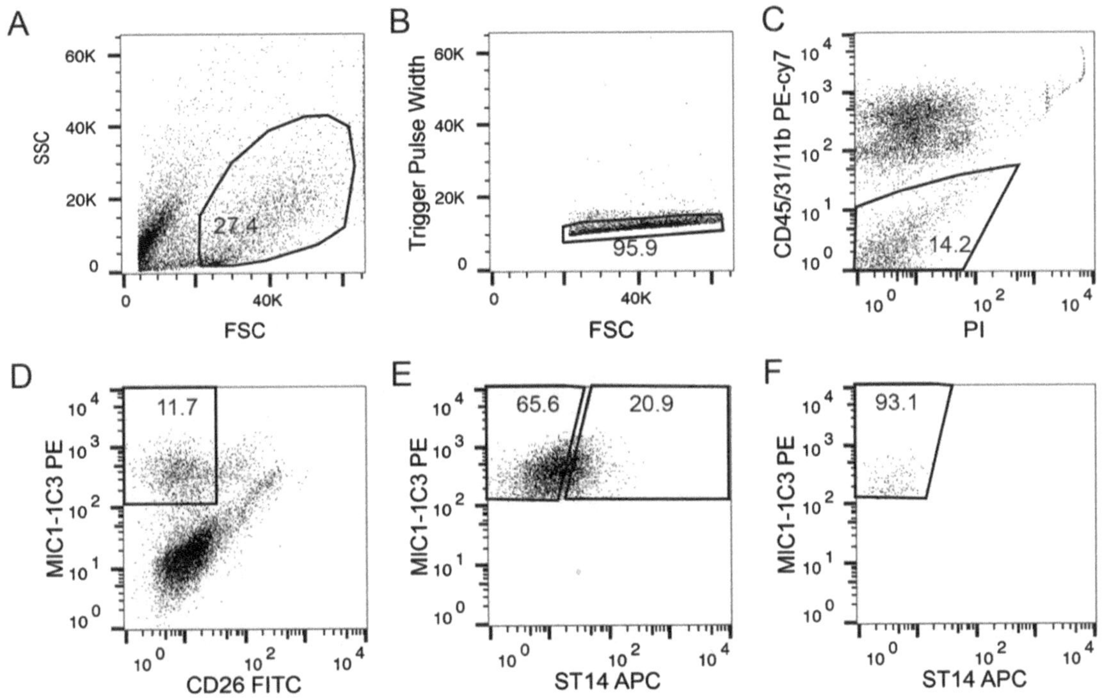

Fig. 3 FACS gating strategy. (**a** and **b**) Cells were sequentially gated based on cell size (forward scatter [FSC] versus side scatter [SSC]). (**a**) Depletion of doublets (FSC vs. trigger pulse width) (**b**). (**c**) Dead cells and debris were excluded by PI positivity. Concurrently a combination of CD45, CD31, and CD11b antibodies was used for depleting blood, endothelium, and Kupffer cells. (**d**) CD26 (DPPIV) was used for hepatocyte staining. (**e**) MIC1-1C3$^+$ cells can be subdivided into two populations: ST14 high (ST14^{hi}M$^+$) and ST14 low (ST14^{lo}M$^+$). (**f**) Fluorescence minus 1 without secondary antibody APC was used to draw gate for ST14 APC positivity and negativity (modified from Bin Li et al., 2017 [3])

11. Last wash and pellet the cells. Resuspend cells with PI at 10 μg/mL in FACS buffer. The FACS gates are drawn as shown in Fig. 3.

3.3 Organoid Cell Culture

1. Precool the 96-well plate, multichannel pipettes, and tips.
2. Make the 5% Matrigel in fresh organoid culture media before sorting, using multichannel pipette to add media into 96-well plate with 100 μL per well. Keep the plate on ice.
3. Sorted cells are directly injected into cold 96-well plate at a density of 1 cell per well.
4. Carefully flip the plate to help the cells sink into the media. The plate will be moved into 37 °C CO_2 incubator immediately.
5. Add more media into organoid culture at day 5. The organoid should be seen clearly at day 7. The organoid-forming assay should be done at day 14.
6. For organoid passage: move the organoid with media in each well into 1.5 mL Eppendorf tube, add the same volume of

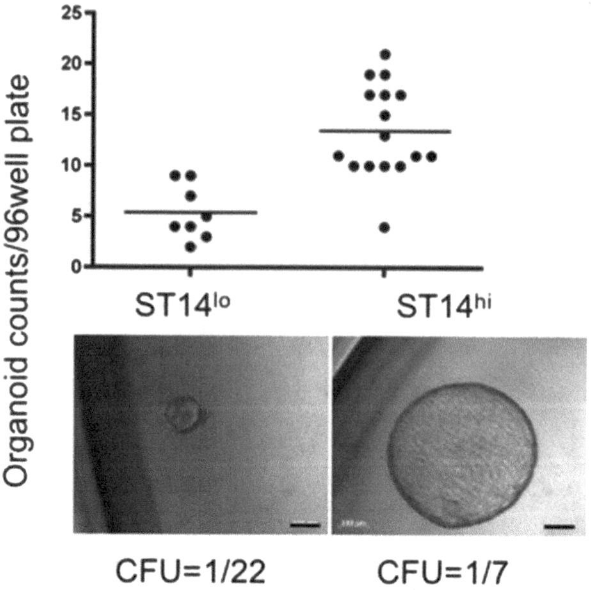

Fig. 4 Representative picture for single-cell-derived organoid under culture day 14. Upper panel: colony-forming efficiency of single-cell-derived colonies in 96 well. Lower panel: representative picture of single $ST14^{hi}M^{+}$ and $ST14^{lo}M^{+}$ cell-derived organoid in 96 well. CFU: colony-forming unit. Scale bar = 100 μm (modified from Bin Li et al., 2017 [3])

TrypLE, pipette with P200, and look through the tube under microscope. Matrigel dissolves in TrypLE. If organoids are still stiff, add more TrypLE at room temperature (RT), or put the tube in 37 °C for 3 min.

7. When dissociated small clumps as well as single cell are seen, add wash medium to wash and spin.
8. The pellet is resuspended in 50 μL Matrigel on ice.
9. Pipette Matrigel droplet on prewarmed 24-well suspension culture plate at 37 °C.
10. After 5 min, add prewarmed organoid medium into plate. The medium will be changed twice a week. After 14 days culture, the organoid forming efficiencies are calculated, *see* Fig. 4.

4 Notes

1. Solutions 1 and 2 are kept at room temperature, and solution 3 is prepared fresh prior to use.
2. Thaw Matrigel overnight on ice.
3. To make sure the whole surgery is at 37 °C, set the water bath to 40 °C.
4. Make sure to add liquid to air trap!

5. If the *portal vein* cannulation failed, cannulate *vena cava* instead.
6. At 4 mL/min, there is about a 1 min lag between switching solutions and them reaching the animal. Once the liver is soft and not bond, it is the time to stop.
7. Tube #1, PI-only control; #2, FITC-only control; #3, PE-only control; #4, PEcy7-only control; #5, APC-only control; #6, florescence minus APC control (second antibody APC only, no primary ST14); #7, florescence minus PE control (second antibody PE only, no primary MIC1-1C3).

Acknowledgments

The authors thank members in the Markus Grompe Lab at Oregon Health and Science University for technical supports. The authors also thank Oregon Health and Science University Flow Cytometry Core for technical assistance.

References

1. Dorrell C et al (2011) Prospective isolation of a bipotential clonogenic liver progenitor cell in adult mice. Genes Dev 25:1193–1203. https://doi.org/10.1101/gad.2029411
2. Azuma H et al (2007) Robust expansion of human hepatocytes in Fah−/−/Rag2−/−/Il2rg−/− mice. Nat Biotechnol 25:903–910. https://doi.org/10.1038/nbt1326
3. Li B et al (2017) Adult mouse liver contains two distinct populations of cholangiocytes. Stem Cell Reports 9:478–489. https://doi.org/10.1016/j.stemcr.2017.06.003
4. Huch M et al (2013) In vitro expansion of single Lgr5+ liver stem cells induced by Wnt-driven regeneration. Nature 494:247–250. https://doi.org/10.1038/nature11826
5. Dorrell C et al (2014) The organoid-initiating cells in mouse pancreas and liver are phenotypically and functionally similar. Stem Cell Res 13:275–283. https://doi.org/10.1016/j.scr.2014.07.006

Chapter 4

Isolation and Expansion of Rat Hepatocytic Progenitor Cells

Junichi Kino, Norihisa Ichinohe, Masayuki Ishii, and Toshihiro Mitaka

Abstract

This protocol showed procedures to isolate and expand small hepatocytes (SHs), as hepatocytic progenitor cells, from a rat liver. SHs are identified as a subpopulation of mature hepatocytes in a healthy liver. SHs can proliferate to form colonies in serum-free medium on hyaluronic acid-coated dishes, of which cells show CD44 positivity ($CD44^+$ SHs). $CD44^+$ SHs can be separated and purified from colonies by using anti-CD44 antibodies after enzymatic dissociation. $CD44^+$ SHs can proliferate to form colonies on Engelbreth-Holm-Swarm gel (EHS-gel)-coated dishes in the serum-free medium for a long period and subculture for several times. Even after the second passage, the cells possess characteristics of hepatocytes such as expression of albumin and HNF4α. In addition, when the cells are treated with EHS-gel, they can recover highly differentiated functions of hepatocytes such as glycogen production, CYP activity, and bile secretion.

Key words Small hepatocyte, Serum-free culture, CD44, Hepatocytic function

1 Introduction

Hepatocytes are the predominant cell type in the liver, constituting over 80% of the liver mass under physiological conditions [1]. The liver can regenerate after surgical resection or cytotoxic damages. Following a 2/3 partial hepatectomy (PH), residual hepatocytes proliferate and replace the lost liver mass [2]. Rat mature hepatocytes (MHs) were shown to perform at least 18 cell doublings [3]. On the other hand, it has been shown that mouse hepatocytes possess tremendous replication potential by using the technique of serial transplantations [4, 5]. However, primary hepatocytes are well known to have poor growth activity *in vitro* and difficult to maintain metabolic and xenobiotic functions [6, 7]. Various attempts have been made over the last several decades to harness the innate replication potential of hepatocytes *ex vivo*. Small

All the authors contributed equally to this work.

Naoki Tanimizu (ed.), *Hepatic Stem Cells: Methods and Protocols*, Methods in Molecular Biology, vol. 1905,
https://doi.org/10.1007/978-1-4939-8961-4_4,

hepatocytes (SHs), hepatocytic progenitor cells, have shown to possess the highest growth potential in culture [8, 9]. SHs were isolated from a healthy adult liver of rat and human and proliferated to form colonies in a serum-free medium on hyaluronic acid (HA)-coated dishes [10, 11], which may be dependent on CD44 expression at the cell surface [12]. Although CD44 expression is weaker than that of a rat, mouse SHs were also identified [13]. The colonies originated from a single cell consisting of approximately 30–40 cells after 8–10 days of plating [10, 12, 14]. Morphology of each SH is similar to MHs except the size. The colonies consisting of small-sized cells can be isolated by using collagenase and hyaluronidase. $CD44^+$ SHs are isolated and sorted by anti-CD44 antibody, and then plated on Engelbreth-Holm-Swarm (EHS)-gel-coated dishes. Attached cells can proliferate to form colonies in the serum-free medium [15]. The cells possessing CD44 positivity and the ability of albumin production can be passaged at least four times. Although the size and other features of most colonies are uniform in the early culture period after the passage, the features of colonies are divided into two types with time in culture; one is consisted of small-sized cells and the other of large and flattened ones. In addition, maturation of SHs can be manually induced by EHS-gel treatment, and the cells recover highly differentiated functions such as glycogen storage, cytochrome P450 (CYP) activity, and bile secretion. This protocol shows the procedures how to isolate rat $CD44^+$ SHs maintaining hepatocytic characteristics and to passage them (Fig. 1).

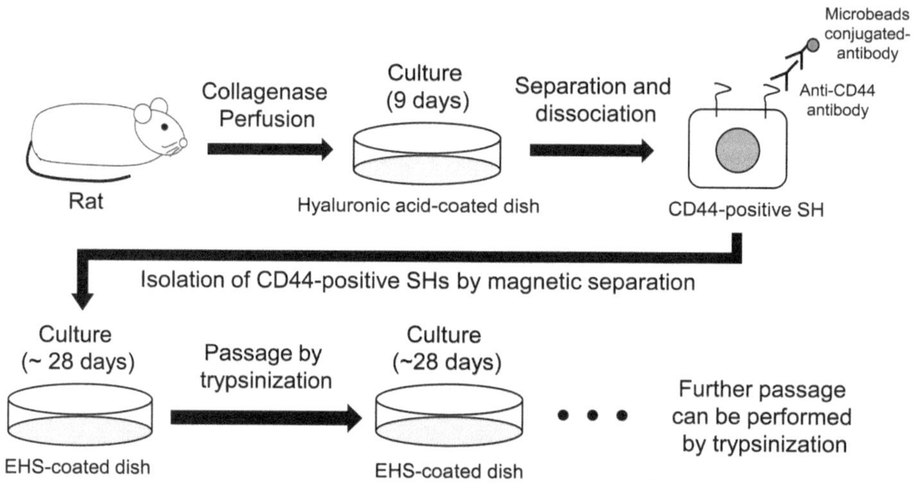

Fig. 1 Overview of procedures for the isolation and expansion of rat small hepatocytes (SHs). Liver cells are isolated from a 6–10 week-old male rat. SHs form colonies on hyaluronic acid-coated dish and express CD44 on their cell surfaces. CD44-positive SHs ($CD44^+$ SHs) is purified by magnetic separation using anti-CD44 antibody. $CD44^+$ SHs can expand on EHS-coated dish and be passaged after trypsinization

2 Materials

2.1 Isolation of Liver Cells

The 6- to 10-week-old F344 rats are used for cell isolation. The animal is maintained with standard chow and water ad libitum. All apparatuses used for the collagenase perfusion are sterilized by autoclave or treatment of 70% ethanol before the experiment starts. All animal experiments must comply with national and institutional regulations.

1. Phosphate-buffered saline without Mg^{2+}/Ca^{2+} (PBS). Store at room temperature (RT).
2. 500 μg/mL insulin stock solution (1000×): Add 100 mg insulin to 100 mL double-distilled water (ddH_2O), and subsequently add 1.2 mL 1 N HCl (*see* **Note 1**). Adjust to 200 mL with ddH_2O and filtrate with a 0.2-μm filter.
3. Pre-perfusion solution: Add approximately 850 mL ddH_2O into a 1000-mL graduated cylinder 100-mL Ca^{2+}, Mg^{2+}-free HANKS' balanced salt solution (HBSS), 190 mg ethylene glycol tetraacetic acid (EGTA), and 1 mL 1000× insulin stock solution and stir, and then adjust pH to 7.5 with 1 M $NaHCO_3$. Adjust to 1000 mL with ddH_2O and filtrate with a 0.2-μm filter. Store at 4 °C until use.
4. Perfusion solution: Add 1 mL 1000× insulin stock solution to 200 mL HBSS. Add collagenase (Wako Pure Chemical Industries) of 26,000 U to the pre-warmed perfusion solution, and make 130 U/mL collagenase solution just before a perfusion (*see* **Note 2**).
5. Wash solution: Add 0.5 mL 1000× insulin stock solution, 2 mL penicillin-streptomycin solution (PC-SM, Sigma), and 0.5 mL gentamicin solution (50 mg/mL, Sigma) to 500 mL HBSS. Store at 4 °C until use.
6. Perfusion apparatus.
7. Water bath.
8. Peristatic pump (RP-1000, Tokyo Rika Instruments).
9. Silicon tube: inside diameter, 4.76 mm; outside diameter, 7.94 mm.
10. O_2 gas (95% O_2 and 5% CO_2).
11. Isoflurane and vaporizer.
12. Butterfly needle, 18 gauge.
13. Surgical scissor.
14. Tweezers.
15. Mosquito forceps.
16. Surgical clip, bulldog type.

17. Surgical thread.
18. Heating lamp.
19. Petri dish (100 mm).
20. Beaker (100 mL).
21. Conical tube (50 mL).
22. Sterilized serological pipette.
23. Pipetting controller.
24. Tabletop centrifuge.
25. Nylon filter net (250 μm).
26. Cell strainer: 70-μm filter (Corning).
27. Neubauer improved hemocytometer.
28. 0.1% trypan blue solution.

2.2 Culture of the Isolated Liver Cells on Hyaluronic Acid (HA)-Coated Dishes

1. 10 mg/mL HA stock solution: Dissolve HA-derived *Streptococcus equi* (SIGMA, Cat No. 53747) in sterilized PBS by stirring using a magnetic stir bar at 37 °C overnight (*see* **Note 3**). Store HA stock solution at 4 °C.
2. Plastic culture dishes, 100, 60, or 35 mm.
3. 10^{-4} M Dexamethasone stock solution: Add 39.2 mg dexamethasone to an autoclaved brown bottle. Add 10 mL absolute ethanol (10^{-2} M stock solution) and then dilute by 100-fold with autoclaved ddH_2O. Store at 4 °C until use.
4. 10 μg/mL Epidermal growth factor (EGF) stock solution: Add 10 mL autoclaved ddH_2O to mouse EGF 100 μg vial (Corning). Dispense each aliquot in a cryogenic tube and store it at −20 °C until use.
5. 1 M Nicotinamide stock solution: Add 12.21 g nicotinamide to 100 mL PBS. Filtrate with a 0.2-μm filter. Store at 4 °C until use.
6. 100 mM Ascorbic acid-2 phosphate (Asc2P) stock solution: Add 2.90 g Asc2P to100 mL PBS, filtrate with a 0.2-μm filter, and store in a autoclaved brown bottle at 4 °C until use.
7. Culture medium stock solution: Add 15.56 g Dulbecco's modified Eagle's medium/Nutrient Mixture Ham F-12 powder (DMEM/F12, Sigma), 1.20 g HEPES, 30 mg L-Proline, PC-SM to a 1000 mL beaker, and ddH_2O up to 1000 mL. Mix the reagents using a magnetic stir bar, add 2.20 g $NaHCO_3$, adjust to pH 7.6 with 1 N NaOH, and filtrate with a 0.2-μm filter. Store at 4 °C until use.
8. Culture medium: Add 1.67 mL BSA solution (30%), 5.50 mL nicotinamide stock solution, 5 mL Asc2P stock solution, 5 mL insulin-transferrin-selenium solution (ITS-X, Thermo Fisher), 0.5 mL EGF stock solution, 0.5 mL dexamethasone stock

solution, and 0.5 mL gentamicin solution in 500 mL culture medium stock solution. Store at 4 °C until use.

9. CO_2 incubator.

2.3 Isolation of CD44-Positive Cells from the Cells on HA-Coated Dishes

1. PBS. Store at RT.
2. EDTA/PBS: PBS containing 0.02% ethylenediaminetetraacetic acid (EDTA). Store at RT.
3. Collagenase. Store at −20 °C until use.
4. 17,500 U/50 μL Hyaluronidase stock solution: Add hyaluronidase to PBS. Adjust to 3500 U/μL, dispense 50 μL aliquot in a cryogenic tube, and store it at −20 °C until use.
5. Separation solution: Add 75 mg collagenase and 50 μL hyaluronidase stock solution to 75 mL HBSS, and warm it at 37 °C just before separation.
6. Magnetic stirrer and stirring bar.
7. MACS® buffer: Add 2 mL 500 mM EDTA solution and 8.3 mL 30% BSA solution to 500 mL PBS. Store at 4 °C until use.
8. Mouse anti-rat CD44 monoclonal antibody (BD biosciences, Cat No. 554869).
9. Microbead-conjugated anti-mouse IgG_{2a+b} (Miltenyi Biotec).
10. MidiMACS™ separator (Miltenyi Biotec).
11. LS column and its plunger (Miltenyi Biotec).
12. Conical tube (15 and 50 mL).
13. Sterilized serological pipette.
14. Pipetting controller.
15. Tabletop centrifuge.

2.4 Culture and Passage of the CD44-Positive Cells on EHS-Gel-Coated Dishes

1. EHS-gel (Matrigel®, Growth factor-reduced, Corning). Store at −20 °C until use.
2. Plastic culture dishes, 100 , 60, or 35 mm.
3. Culture medium: the same culture medium mentioned above (*see* Subheading 2.2).
4. PBS. Store at RT.
5. PBS containing 0.25% trypsin and 0.01% EDTA. Store at 4 °C until use.
6. Conical tube (15 and 50 mL).
7. Sterilized serological pipette.
8. Pipetting controller.
9. Tabletop centrifuge.
10. Neubauer improved hemocytometer.

11. 0.1% trypan blue solution.
12. CO_2 incubator.

3 Methods

3.1 Isolation of Liver Cells

1. Set up the perfusion apparatus in the warmed water bath at approximately 39 °C. Circulate the pre-perfusion solution in the perfusion apparatus using the peristatic pump and silicone tube before the experiment, and bubble in it with O_2 gas at a flow rate of 0.5 L/min.
2. Anesthetize a rat with an inhalation of isoflurane.
3. Cut the abdominal wall with surgical scissors and open the abdominal cavity.
4. Keep opening the abdominal cavity by grasping the edge of cut abdominal wall using mosquito forceps.
5. Move the intestine to expose the liver, portal vein, bile duct, and splenic vein.
6. Ligate a common bile duct and splenic vein with a surgical thread at the portion nearest the portal vein (*see* **Note 4**).
7. Stop the circulation of the pre-perfusion solution and insert a butterfly needle filled with pre-perfusion solution into the portal vein at 1.5–2.0 cm from the bifurcation of the portal vein. Stop the tip of the needle at a position close to bifurcation of the portal vein, and clamp the needle with a surgical clip.
8. Start the perfusion of the pre-perfusion solution at a flow rate of 30 mL/min (*see* **Note 5**).
9. Cut the inferior vena cava and heart just after starting the perfusion (*see* **Note 6**).
10. When the volume of the pre-perfusion solution becomes small, add the collagenase to the perfusion solution and then pour it into the perfusion apparatus.
11. Flow the solution at a flow rate of 15 mL/min. Heat around the abdominal cavity using a heating lamp during the perfusion (*see* **Note 7**).
12. When the volume of the perfusion solution becomes about half, change the flow rate at 10mL/min (*see* **Note 8**).
13. Stop the flow before air bubbles move into the liver when the solution flows out from the perfusion apparatus.
14. Cut the liver from the abdominal cavity and transfer it to a sterilized Petri dish (*see* **Note 9**).
15. Prepare a 100-mL beaker with 80-mL wash solution.

16. Peel the hepatic capsule carefully, and then, to drop the digested cells, gently shake the liver in the beaker. Dropped cells in the Petri dish are also poured to the beaker.
17. Filter the cell suspension through a 250-μm nylon filter net into a new 100-ml beaker.
18. Filter the cell suspension through a cell strainer with 70-μm filter into four 50-ml conical tubes using a 25-ml serological pipette, and then adjust each tube to an equal volume (approximately 40 mL) with wash solution.
19. Centrifuge the tubes at 50 × *g* for 1 min at 4 °C.
20. Collect the supernatants and transfer to four new 50-mL conical tubes (*see* **Note 10**), and wash solution is then added to each tube up to 45 mL. Repeat **Steps 19** and **20** twice.
21. Centrifuge the tubes at 45 × *g* for 5 min at 4 °C (*see* **Note 11**).
22. Aspirate the supernatant of 40 mL, then add new 40-mL wash solution to each tube, and dissociate the cell pellet by pipetting.
23. Centrifuge the tubes at 150 × *g* for 5 min at 4 °C (*see* **Note 12**).
24. Repeat **Steps 22** and **23** twice. Aspirate the supernatant as possible, and then add new 20-mL wash solution to each tube, and dissociate the cell pellet by pipetting.
25. Gather suspension into two 50-mL conical tubes, and centrifuge the tubes at 150 × *g* for 5 min at 4 °C.
26. Aspirate the supernatant as possible, and then add new 20-mL culture medium to each tube, and dissociate the cell pellet by pipetting.
27. Gather the suspension into one 50-mL conical tube, and centrifuge the tube at 50 × *g* for 5 min at 4 °C.
28. Aspirate the supernatant as possible, and then add new 10-mL culture medium to the tube, and dissociate the cell pellet by pipetting. Put this tube on ice.

3.2 Culture of the Isolated Cells on HA-Coated Dishes

1. Dilute the HA stock solution to 1 mg/mL with PBS, fill plastic culture dishes, and incubate overnight at 37 °C. On the next day, aspirate the HA solution and wash once with PBS. Aspirate PBS and dry in a clean bench for more than 30 min.
2. Mix 0.5 mL of the suspension of the isolated cells and 1.5 mL trypan blue solution.
3. Count the number of viable and dead cells, which are smaller than MHs and larger than nonparenchymal cells, using Neubauer improved hemocytometer (*see* **Note 13**).
4. Adjust cell density by diluting the suspension with an appropriate volume of the culture medium.

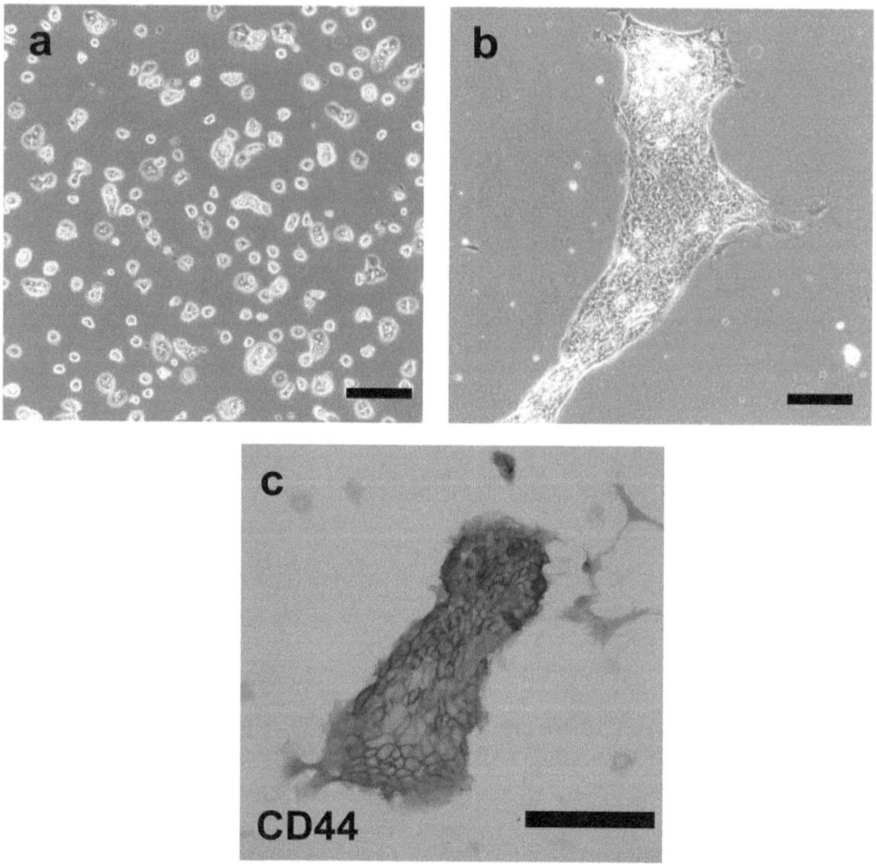

Fig. 2 Formation of SH colonies on hyaluronic acid (HA)-coated dish. (**a**) One day after seeding, SHs adhere on HA-coated dish. (**b**) SHs proliferate to form colonies at 9 days after plating, of which consist small-sized cells. (**c**) All cells in colonies express CD44 on their cell surfaces. Scale bar: 200 μm

5. Wash dry HA-coated dishes with PBS just before cell seeding.
6. Seed the cells (2.0×10^4 cells/cm^2) on the HA-coated dishes.
7. Place the dishes in 5% CO_2/95% air-incubator at 37 °C for 3 h.
8. Replace the medium with fresh medium (*see* **Note 14**).
9. Place the dishes in 5% CO_2/95% air-incubator at 37 °C for 9 days, and replace the medium every other day (*see* Fig. 2 and **Note 15**).

3.3 Isolation of CD44-Positive Cells from the Cells on HA-Coated Dishes

1. Wash the SH colonies on the HA-coated dishes with PBS.
2. Treat the colonies with 0.02% EDTA/PBS for 5 min (*see* **Note 16**).
3. After aspiration of the solution, treat the colonies with the separation solution for 5 min at 37 °C.

4. Collect the detached cells from HA-coated dishes into a sterilized 100-mL beaker.
5. Stir the cell suspension using a magnetic stirrer and stir bar for 30 min at 37 °C.
6. Collect the cell suspension and centrifuge it at 150 × *g* for 5 min at 4 °C.
7. Aspirate the supernatant and suspend the cell pellet in 20 mL MACS buffer (*see* **Note 17**).
8. Centrifuge the cell suspension at 150 × *g* for 5 min at 4 °C.
9. Aspirate the supernatant and suspend the cell pellet in 1-mL mouse anti-rat CD44 antibody solution (1:1000 dilution, diluted in MACS buffer; *see* **Note 18**).
10. Incubated the cell suspension for 60 min on ice.
11. Add 20 mL MACS buffer to the cell suspension and centrifuge it at 150 × *g* for 5 min at 4 °C.
12. Aspirate the supernatant and suspend the cell pellet in 1 mL microbead-conjugated anti-mouse IgG_{2a+b} solution (1:5 dilution, diluted in MACS buffer; *see* **Note 19**).
13. Incubate the cell suspension for 30 min on ice.
14. Add 20 mL MACS buffer to the cell suspension and centrifuge it at 150 × *g* for 5 min at 4 °C.
15. Aspirate the supernatant and suspend the cell pellet in 2 mL MACS buffer.
16. Install LS column on a MidiMACS separator, and set a 50-mL conical tube under the column.
17. Apply 2 mL MACS buffer onto the column.
18. Apply the cell suspension onto the column after the completion of dripping the filtrate from the column (*see* **Note 20**).
19. Apply 2 mL MACS buffer onto the column after the completion of dripping the filtrate from the column twice.
20. Remove the column from the MidiMACS separator after the completion of dripping the filtrate from the column.
21. Place the column on the 15-mL conical tube.
22. Add 5 mL MACS buffer to the column and immediately flush out a fraction of CD44-positive cells using the plunger.

3.4 Coating Dishes with EHS-Gel

1. Dispense each aliquot of EHS-gel in a cryogenic tube and store it at −20 °C until use.
2. Thaw EHS-gel stock solution on ice.
3. Dilute EHS-gel with HBSS on ice and adjust the concentration to 0.2 mg protein/mL (*see* **Note 21**).

4. Pour the diluted EHS-gel solution into each plastic culture dish (1 mL/35-mm, 2 mL/60-mm, 5 mL/100-mm dish).
5. Incubate the dishes at 37 °C overnight.
6. Aspirate the solution in the dishes and dry them in a clean bench.
7. Sterilize the dishes with UV irradiation for about 3 min.

3.5 Culture of the CD44-Positive ($CD44^+$) Cells on the EHS-Gel-Coated Dishes

1. Mix 1 volume of the suspension of $CD44^+$ cells and 3 volumes of trypan blue solution.
2. Count the number of viable and dead cells using a Neubauer improved hemocytometer.
3. Centrifuge the cell suspension at 150 × *g* for 5 min at 4 °C.
4. Aspirate the supernatant and suspend the cell pellet in an appropriate volume of the culture medium for adjusting cell density (*see* **Note 22**).
5. Wash dry EHS-gel-coated dishes with PBS just before cell seeding.
6. Seed the cells on the EHS-gel-coated dishes.
7. Place the dishes in 5% CO_2/95% air-incubator at 37 °C for 3 h.
8. Replace the medium with fresh culture medium.
9. Place the dishes in 5% CO_2/95% air-incubator at 37 °C, and replace the medium every other day (*see* Fig. 3 and **Note 23**).

3.6 Passage of the $CD44^+$ Cells on the EHS-Gel-Coated Dishes

1. Wash the $CD44^+$ cells on the EHS-gel-coated dishes with PBS.
2. Incubate the cells with PBS containing 0.25% trypsin and 0.01% EDTA in 5% CO_2/95% air-incubator at 37 °C for 10 min.
3. Dissociate the cells by pipetting gently.
4. Add the culture medium to the cell suspension to stop the digestion by trypsin.
5. Centrifuge the cell suspension at 150 × *g* for 5 min at 4 °C.

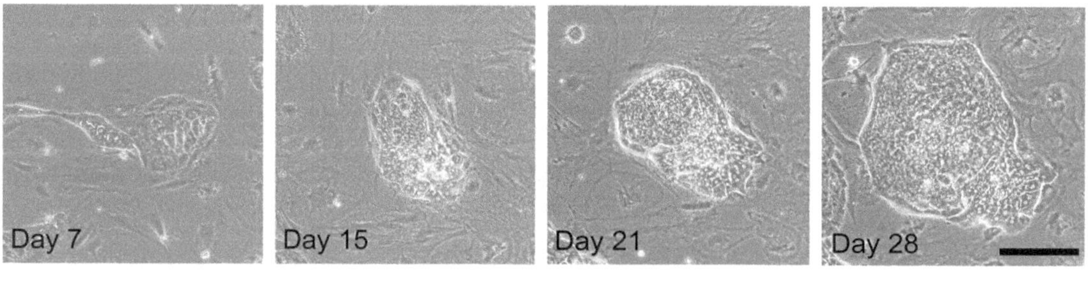

Fig. 3 Growth of sorted $CD44^+$ SHs on an EHS-gel-coated dish. Attached $CD44^+$ SHs can grow to form colonies on EHS-gel-coated dishes. The same colony was followed until 28 days after the passage. All photos show the same magnification. Scale bars; 200 μm

6. Aspirate the supernatant and suspend the cell pellet in the culture medium.
7. Mix 1 volume of the suspension of $CD44^+$ cells and 3 volumes of trypan blue solution.
8. Count the number of viable and dead cells using a Neubauer improved hemocytometer.
9. Wash dry EHS-gel-coated dishes with PBS just before cell seeding.
10. Dilute the cell suspension with an appropriate volume of the culture medium for adjusting cell density (*see* **Note 22**).
11. Seed the cells on the EHS-gel-coated dishes.

4 Notes

1. Insulin is dissolved under acidic condition.
2. Collagenase must be dissolved just before the perfusion and not be shaken vigorously, because its activity is easily decreased.
3. HA should be sterilized with UV irradiation. HA is not dissolved easily in PBS at room temperature; therefore, HA-stock solution should be made until 1 day before coating culture dishes with HA.
4. If the ligation is loose, it may cause an inadequate digestion by collagenase.
5. Pre-perfusion is performed for blood removal and calcium chelating by EGTA in the liver. This step needs for a good collagenase digestion.
6. The perfusate is flowed out from the cut ends of inferior vena cava. Be careful not to store the perfusate in the abdominal cavities. Retention of the collagenase solution causes the early rupture of hepatic capsules.
7. Heating by a lamp makes collagenase digestion in the liver better, because collagenase activity is the highest at 37 °C.
8. Collagenase can digest the liver tissue well by reducing the flow rate of perfusion.
9. The liver becomes soft by the collagenase digestion. Be careful not to disrupt the hepatic capsule during removing the tissues adjacent to the liver.
10. MHs have higher density than SHs and nonparenchymal cells, so MHs are removed from the cell suspension by low-speed and short-time centrifuge.
11. This centrifuge is carried out to remove hematopoietic cells.

12. This centrifuge is carried out to damage the remaining MHs in the cell suspension.
13. Count the number of cells as soon as trypan blue solution is added. Dead cells have nuclei stained with trypan blue. The viability of cells is about 80–90% when the isolation procedure is performed successfully. The lower viability may be caused by an inadequate digestion by collagenase. In this case, modify the perfusion procedure, e.g., the collagenase activity in the perfusion solution or the time of collagenase digestion by changing the flow rate of perfusion.
14. Before the medium change, check the adhesion of cells on the surface of dishes. If the number of attached cells is few, keep the dishes in the incubator at 37 °C for additional 1–2 h. The adhesion of cells on HA-coated cells is weak on the day of seeding, so the medium change is performed carefully.
15. During 9 days of culture, SHs formed colonies consisting of more than 30 cells. When the colonies of SHs are cultured for a time longer than 9 days, adhesion of colonies on the dishes becomes weaker.
16. The cells will be detached from the dishes when treated with 0.02% EDTA/PBS, if the colonies weakly adhere on the dishes. In this case, collect the detached cells in EDTA/PBS and centrifuge it at 150 × *g* for 5 min at 4 °C. The pellet is used for the subsequent treatment of collagenase (**Step 5**).
17. After the treatment of collagenase, the procedures should be performed on ice.
18. One milliliter mouse anti-rat CD44 antibody solution affords to bind 1×10^8 cells. If the total cell number is more than this number, increase the volume of mouse anti-rat CD44 antibody solution.
19. One milliliter microbead-conjugated anti-mouse IgG_{2a+b} solution affords to bind 1×10^8 cells. If the total cell number is more than this number, increase the volume of mouse anti-rat CD44 antibody solution.
20. The column can clog with cells in case that the number of cells is too many. Pipette the suspension above the column in order to remove the clogging.
21. EHS-gel forms a gel above 10 °C. Use pre-cooled pipets, tips, and tubes when handling EHS-gel.
22. SH growth speed depends on the cell density, the number of viable cells, which have the potential to attach to the dishes. The cell density should be more than 1.0×10^4 cells/cm^2 for expanding SHs.

23. The longer a culture period is, the harder the connection between cells become. The culture period should not be longer than 4 weeks, if the cells will be passaged (*see* Subheading 3.6).

References

1. Saxena R, Zucker SD, Crawford JM (2003) Anatomy and physiology of the liver. In: Zakim D, Boyer TD (eds) Hepatology, 4th edn. Saunders, Philadelphia, pp 3–30
2. Higgins GM, Anderson RM (1931) Experimental pathology of the liver. I: Restoration of the liver of the white rat following partial surgical removal. Arch Pathol 12:186–202
3. Stöcker E, Heine WD (1971) Regeneration of liver parenchyma under normal and pathological conditions. Beitr Pathol 144:400–408
4. Overturf K, al-Dhalimy M, Ou CN, Finegold M, Grompe M (1997) Serial transplantation reveals the stem-cell-like regenerative potential of adult mouse hepatocytes. Am J Pathol 151:1273–1280
5. Wang MJ, Chen F, Li JX, Liu CC, Zhang HB, Xia Y et al (2014) Reversal of hepatocyte senescence after continuous in vivo cell proliferation. Hepatology 60:349–361
6. Bhatia SN, Underhill GH, Zaret KS, Fox IJ (2014) Cell and tissue engineering for liver disease. Sci Transl Med 6:245sr2
7. Mitaka T (1998) The current status of primary hepatocyte culture. Int J Exp Pathol 79:393–409
8. Mitaka T, Mikami M, Sattler GL, Pitot HC, Mochizuki Y (1992) Small cell colonies appear in the primary culture of adult rat hepatocytes in the presence of nicotinamide and epidermal growth factor. Hepatology 16:440–447
9. Mitaka T, Kojima T, Mizuguchi T, Mochizuki Y (1995) Growth and maturation of small hepatocytes isolated from adult rat liver. Biochem Biophys Res Commun 214:310–317
10. Chen Q, Kon J, Ooe H, Sasaki K, Mitaka T (2007) Selective proliferation of rat hepatocyte progenitor cells in serum-free culture. Nat Protoc 2:1197–1205
11. Sasaki K, Kon J, Mizuguchi T, Chen Q, Ooe H, Oshima H et al (2008) Proliferation of hepatocyte progenitor cells isolated from adult human livers in serum-free medium. Cell Transplant 17:1221–1230
12. Kon J, Ooe H, Oshima H, Kikkawa Y, Mitaka T (2006) Expression of CD44 in rat hepatic progenitor cells. J Hepatol 45:90–98
13. Tanimizu N, Ichinohe N, Ishii M, Kino J, Mizuguchi T, Hirata K, Mitaka T (2016) Liver progenitors isolated from adult healthy mouse liver efficiently differentiate to functional hepatocytes in vitro and repopulate liver tissue. Stem Cells 35:2889–2901
14. Mitaka T, Sato F, Mizuguchi T, Yokono T, Mochizuki Y (1999) Reconstruction of hepatic organoid by rat small hepatocytes and hepatic nonparenchymal cells. Hepatology 29:111–125
15. Ishii M, Kino J, Ichinohe N, Tanimizu N, Ninomiya T, Suzuki H et al (2017) Hepatocytic parental progenitor cells of rat small hepatocytes maintain self-renewal capability after long-term culture. Sci Rep 7:46177

Part II

Characterization of Liver Progenitors In Vivo

Chapter 5

Genetic Lineage Tracing of Biliary Epithelial Cells

Teresa Rubio-Tomás, Beatriz Aguilar-Bravo, and Pau Sancho-Bru

Abstract

Lineage tracing of liver cells is a powerful tool to understand liver embryonic development, healthy liver cell homeostasis, tissue repair, and regeneration. Lineage tracing of biliary epithelial cells (BECs) in the adult liver has been used to assess the contribution of the biliary epithelium to liver injury, regeneration, and disease. These studies have shown the contribution of BECs to the expansion of ductular reaction (DR) and liver progenitor cells (LPCs) and eventually the generation of new hepatocytes. Few genetic lineage-tracing mouse models have been proved to trace BECs. This chapter is focused on lineage tracing of BECs in mouse models of liver injury and regeneration. First, we mention different existing approaches to trace the biliary epithelium based on proteins specifically expressed by BECs such as sex-determining region Y-box 9 (SOX9), osteopontin (OPN), and cytokeratin-19 (KRT19). Second, we describe mouse models that can be used to evaluate cell fate during liver injury and regeneration (i.e., partial hepatectomy (PHx), acute liver injury models, and chronic liver damage models such as 3,5-diethoxycarbonyl-1,4-dihydro-collidin (DDC) diet, choline-deficient ethionine-supplemented (CDE) diet, or chronic carbon tetrachloride (CCl_4) administration). Third, we suggest possible readouts to assess BECs fate based on immunofluorescence analysis.

Key words Lineage tracing, Liver regeneration, Biliary epithelial cells, Liver progenitor cells, Ductular reaction, Chronic liver injury, Animal models

1 Introduction

In the last years, development of lineage-tracing technology has allowed the possibility of better understanding cell fate in liver development as well as in the context of liver injury and regeneration, tissue repair, and cancer. Besides its potential to investigate stem cell commitment and differentiation, the use of lineage tracing in adult organs with regenerative capacity has provided important information to understand tissue cell homeostasis, regeneration, and disease. In healthy condition the liver is a slow-cycling organ; however, after liver damage resulting from diverse insults, loss of liver mass, toxicity, or diseases, the liver rapidly regenerates recovering the liver cell mass and function. When hepatocyte proliferation is intact, such as in the model of partial hepatectomy

Naoki Tanimizu (ed.), *Hepatic Stem Cells: Methods and Protocols*, Methods in Molecular Biology, vol. 1905, https://doi.org/10.1007/978-1-4939-8961-4_5, © Springer Science+Business Media, LLC, part of Springer Nature 2019

(PHx), liver regeneration and recovery of the cell mass is due to mature cell proliferation, comprising proliferation of hepatocytes and cholangiocytes but also other non-parenchymal cells. However, when hepatocyte regeneration and replication is compromised such as in severe liver injury or in the context of chronic liver diseases, biliary epithelial cells (BECs) can undergo substantial morphological, molecular, and functional changes giving rise to the ductular reaction (DR). This is particularly relevant under situations comprising biliary injury such as biliary diseases or in highly cholestatic advanced chronic liver diseases such as alcoholic cirrhosis and alcoholic hepatitis, in which an important biliary reaction can be observed [1]. DR is a heterogeneous population of cells comprising reactive cholangiocytes as well as more immature cells with a liver progenitor cell (LPC) phenotype characterized by the expression of both hepatocyte and biliary markers as well as stem cell markers [2, 3].

Liver cells have been shown to be highly plastic. Several studies using lineage-tracing strategies have shown the conversion of biliary cells to mature hepatocytes, as well as the transdifferentiation of hepatocytes into biliary cells. Moreover, both cell types have been shown to be able to generate facultative LPCs with the capacity to proliferate and to differentiate into hepatocyte and cholangiocytes [4–7]. LPCs have not been identified in the healthy liver, but they appear and expand after injury, when hepatocytes regenerative capacity is compromised. Huch et al. used Lgr5-LacZ mice to prove that this population expresses the stem cell marker *Lgr5* [8]. LPCs arise in several chronic liver diseases and animal models, and its expansion in the parenchymal region correlates with the degree of liver injury and progression of the disease [9, 10]. LPCs stain positively for BECs markers and often form clusters with duct-like and/or cord-like structures.

The important plasticity of the liver and the phenotypic and functional changes of the BECs reinforce the importance of lineage-tracing strategies to investigate the origin and fate of cells of the DR and to understand the contribution of the biliary epithelium to liver regeneration. Moreover, it can be a powerful tool to investigate the molecular mechanisms underlying the development of DR and the generation of facultative LPCs [11].

Few genetic lineage-tracing systems have been developed to trace BECs. Carpentier et al. used sex-determining region Y-box 9 (SOX9)-CreERT to trace embryonic ductal epithelium cells. They found that these cells gave rise to cholangiocytes lining all segments of the intrahepatic bile ducts, as well as periportal hepatocytes [12]. Español-Suñer et al. showed that osteopontin (OPN)-labeled adult liver cells have the potential to give rise to transit-amplifying progenitor cells and mature hepatocytes in mouse chronic injury models [13]. Lineage-tracing studies performed with the BEC marker cytokeratin-19 (KRT19) have shown different outcomes depending on the mouse model that was used. In most animal models (i.e., 3,5-diethoxycarbonyl-1,4-dihydrocollidin [DDC] diet, choline-deficient ethionine-supplemented [CDE] diet, partial hepatectomy [PHx] and carbon tetrachloride [CCl_4]), there was no evidence that BECs and other non-hepatocyte populations contribute significantly to hepatocyte neogenesis. However, more stringent models inducing hepatocyte cell arrest showed the generation of hepatocytes from cholangiocytes [14, 15]. Moreover, lineage-tracing models are also powerful tools for studying cancer development. Guest et al. used KRT19-CreER^T to target p53 loss in cholangiocytes and observed the appearance of labeled intrahepatic cholangiocarcinoma in response to chronic injury [16]. In Rodrigo-Torres et al., we used the hepatocyte nuclear factor 1-beta (HNF1β) lineage-tracing animal model to trace the biliary compartment under chronic injury [17]. In this study we showed the BECs expressing HNF1β give rise to the DR. However, under the animal models tested, the contribution of HNF1β cells to newly generated hepatocytes was absent or minimal. The procedures described in this protocol are based on this study.

Unfortunately lineage-tracing studies have limitations. First, lack of Cre expression in some target cells could lead to difficulties, especially when reduced populations are targeted. Second, Cre expression in non-targeted cell types can complicate its interpretation and lead to controversy. In order to overcome the limitations due to poor specificity, recent studies have proposed more robust genetic lineage-tracing system that incorporates double recombination based on Dre–rox system to minimize the lack of Cre specificity [18]. Besides intrinsic technical limitations, it is important to mention that animal models do not fully reproduce the complexity of chronic liver diseases and, therefore, results obtained from lineage-tracing studies have to be taken cautiously when translating the findings into human disease.

2 Materials

All chemicals and reagents are commercially available products of analytical grade.

2.1 Mouse Strains

2.1.1 Cre-Expressing Lines

Mice expressing Cre recombinase under the control of a cell-specific promoter can be used to induce site-specific recombination between two loxP sites (Fig. 1). The Cre-loxP system can be used to generate conditional knockout mice by excising part of a protein-coding sequence and, therefore, inhibiting its expression. Cre recombinase sequence can be fused to the estrogen receptor sequence (CreER), and, therefore, Cre recombinase activity becomes tamoxifen-inducible. The use of inducible CreER is particularly interesting in the context of lineage tracing, since expression of markers may vary along embryonic development, which

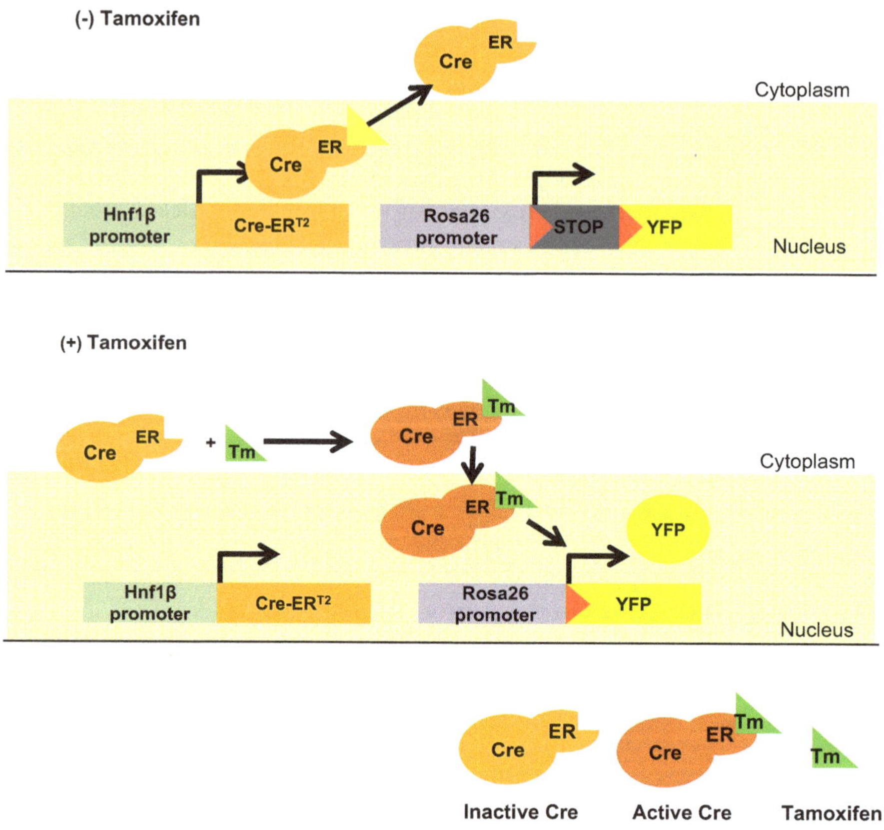

Fig. 1 Tamoxifen-inducible Cre-loxP system. Cre recombinase sequence is fused to hormone-binding domain of the estrogen receptor (CreER). CreER expression is under the control of a biliary epithelial cell (BEC)-specific promoter (e.g., hepatocyte nuclear factor 1-beta [Hnf1β] promoter). In the absence of tamoxifen, CreER remains inactive in the cytoplasm. In the presence of tamoxifen, CreER activates and translocates to the nucleus, where it recombines the region flanked by LoxP sites (red triangles). After the release of the stop codon, yellow fluorescent protein (YFP) is permanently expressed in the cell and its progeny

may lead to confounding results. Different Cre mouse lines can be used for lineage tracing of BEC cells: Hnf1β-CreER (027681, The Jackson Laboratory, United States of America) [17, 19]; SOX9-CreER^{T2} mice (018829, The Jackson Laboratory, United States of America) [12, 20]; OPN-iCreER^{T2} mice [13] and KRT19-CreERT mice (026925, The Jackson Laboratory, United States of America) [15, 21].

2.1.2 Reporter Mouse Lines

In order to generate lineage-tracing animals, cell-specific CreER mice should be crossed with mice bearing Cre-inducible Rosa26R reporters such as LacZ (the gene that codifies for β-galactosidase [β-GAL] protein) or yellow fluorescent protein (YFP), among others (*see* **Note 1**). Lineage-tracing mice are based on the presence of a stop codon flanked by LoxP sites located between the promoter and the reporter gene. When Cre recombinase is expressed, the stop codon is excised and the protein sequence located downstream is expressed. When inducible CreER is used, mice will need to be treated with tamoxifen to induce Cre-mediated LoxP site recombination. Several mouse reporter lines are available, among others: $ROSA26R^{YFP}$ mice (006148, The Jackson Laboratory, United States of America); $ROSA26R^{lacZ}$ mice (002073, The Jackson Laboratory, United States of America) and Gt(ROSA)26Sortm4 (ACTB-tdTomato,-EGFP)Luo (007676, The Jackson Laboratory, United States of America).

2.2 Liver Injury and Regeneration Models

1. Tamoxifen.
2. Corn oil.
3. Surgical instruments, reagents, and equipment required for PHx has been previously described [22].
4. 500 mg/kg mice body weight acetaminophen (APAP).
5. 0.5 mL/kg mice body weight CCl_4. CCl_4 is diluted 1:4 in corn oil.
6. Standard rodent chow diet containing 0.1% DDC.
7. Standard chow diet.
8. Choline-deficient (MP Biomedicals, United States of America), ethionine-supplemented (0.15% in water) (CDE) diet.

2.3 Isolation of BEC-Derived Cells

1. EGTA solution: 137 mM NaCl, 5.4 mM KCl, 0.64 mM $NaH_2PO_4H_2O$, 0.85 mM Na_2HPO_4, 1 mM HEPES, 4.17 mM $NaHCO_3$, 0.5 mM EGTA, and 5 mM glucose, pH = 7.4.
2. Collagenase solution: 0.5 g/L Collagenase A in 137 mM NaCl, 5.4 mM KCl, 0.64 mM $NaH_2PO_4H_2O$, 0.85 mM Na_2HPO_4, 1 mM HEPES, 4.17 mM $NaHCO_3$, and 3.8 mM $CaCl_2{\cdot}2H_2O$, pH = 7.4.
3. Collagenase A (Roche Diagnostics GmbH, Germany).
4. Hank's Balanced Salt Solution (HBSS) (Biological Industries, United States of America).
5. 70-μm cell strainer (Miltenyi Biotec, Germany).
6. Pronase (Roche Diagnostics GmbH, Germany).
7. DNase I (Roche Diagnostics GmbH, Germany).
8. Live/Dead cell stain kit (Life Technologies, United States of America).
9. RLT buffer (Qiagen GmbH, Germany).

2.4 Histological and Biochemical Analysis

Histochemical procedures, immunostaining, colocalization studies, and serum analysis are performed using the following materials:

1. Antibodies against BEC markers: anti-KRT19 (Developmental Studies Hybridoma Bank, University of Iowa, United States of America), anti-epithelial cell adhesion molecule (EpCAM)

(M0804, DakoCytomation, United States of America, and 552370, BD Pharmingen, United States of America), anti-HNF1β (Sc-22840 and Sc-7411, Santa Cruz, United States of America), and anti-SOX9 (AB5535, Millipore Merck, United States of America).

2. Antibodies against hepatocytes, such as anti-hepatocyte nuclear factor 4-alpha (HNF4α) (ab181604, Abcam, United Kingdom).
3. Antibodies against neutrophil markers, such as anti-myeloperoxidase (MPO) (ab9535, Abcam, United Kingdom).
4. Antibodies against macrophage markers, such as anti-F4/80 receptor (MCA497-R, Serotec, United States of America).
5. Antibodies against reporter proteins: anti-β-GAL (A11132, Invitrogen and ab9361, Abcam, United Kingdom), anti-CRE [23], and anti-green fluorescent protein (GFP) (ab6673, Abcam, United Kingdom, and A6455, Invitrogen, United States of America) (*see* **Note 2**).
6. Kits for detection of apoptosis: commercial TUNEL kit (12156792910—In Situ Cell Death Detection Kit, TMR red, Merck, United States of America) and caspase activity kit (G8091, Caspase-Glo 3/7 Assay, Promega, United States of America).
7. Kits for detection of alanine aminotransferase (ALT), aspartate aminotransferase (AST), lactate dehydrogenase (LDH), and alkaline phosphatase (AP) serum levels (ADVIA 1800 Clinical Chemistry System, Siemens Healthcare, Germany).

3 Methods

3.1 Induction of Expression of the Reporter Proteins by Tamoxifen

Tamoxifen is administered to induce Cre-mediated recombination.

1. Dissolve tamoxifen at 100 mg/mL in corn oil (0.9% NaCl and 10% absolute ethanol to facilitate sonication).
2. Treat mice with tamoxifen intraperitoneally every 24 h at a dose of 75 mg/kg for 5 consecutive days (*see* **Note 3**).
3. All animal models of liver injury were started 1 week after the last tamoxifen treatment.

3.2 Induction of Liver Injury and Regeneration

3.2.1 Partial Hepatectomy

Two-thirds PHx in rodents is a well-established model to study liver regeneration (Fig. 2a). Recovery of the liver cell mass after PHx is mediated by processes involving mature cell proliferation, without extensive necrosis or inflammation. Moreover, as the regeneration process starts immediately after PHx, this model can be precisely controlled. As shown in Rodrigo-Torres et al., after PHx there was no contribution of HNF1β cells to hepatocytes.

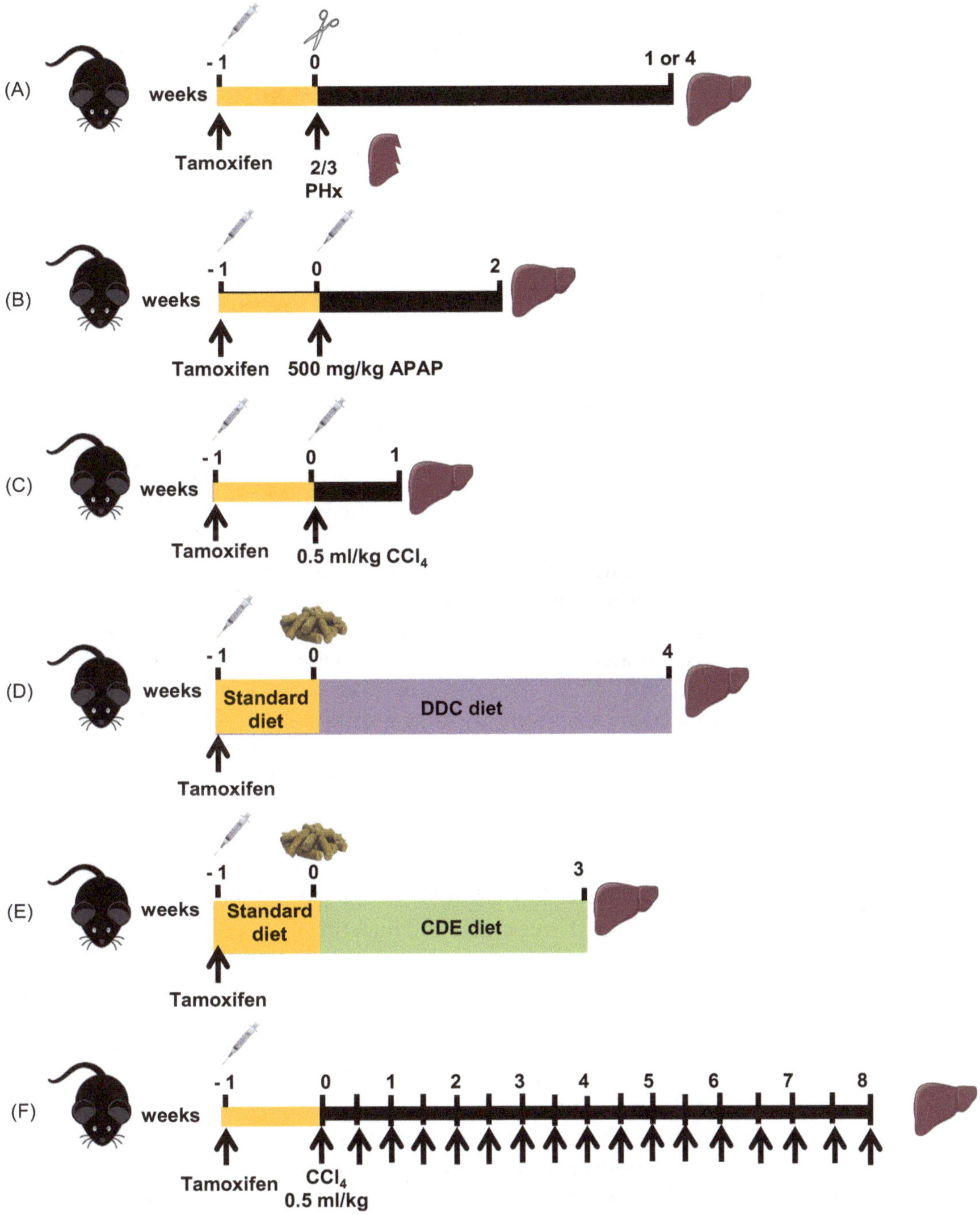

Fig. 2 Experimental models of liver regeneration. (**a**) Partial hepatectomy (PHx), (**b**) acute acetaminophen (APAP) treatment, (**c**) acute carbon tetrachloride (CCl_4) administration, (**d**) 3,5-diethoxycarbonyl-1,4-dihydrocollidin (DDC) diet, (**e**) choline-deficient ethionine-supplemented (CDE) diet, and (**f**) chronic CCl_4 administration

1. Anesthetize mice.
2. Perform two-thirds PHx in mice as previously described [22].
3. Sacrifice 1 week after PHx.

3.2.2 Acute APAP Treatment

APAP model (Fig. 2b) induces hepatocellular death and release of danger-associated molecular patterns (DAMPs) leading to macrophages and NKT cells infiltration. This model mimics drug-induced injury. Mice treated with APAP show a mild induction of DR expansion in the liver. In Rodrigo-Torres et al., no BEC-derived hepatocyte generation was observed when assessed by dual staining of HNF4α/YFP.

1. Fast mice for 8 h with free access to water.
2. Inject intraperitoneally a single dose of 500 mg/kg body weight APAP diluted in DPBS at 35–40 °C.
3. Inject control mice with equivalent amount of DPBS.
4. Sacrifice mice 2 weeks after the injection.

3.2.3 Acute CCl_4 Treatment

Acute treatment with CCl_4 (Fig. 2c) is a widely used model of acute liver failure caused by toxic substances. CCl_4 induces inflammation, thrichloromethyl radical formation, and production of reactive oxygen species (ROS). This leads to lipid peroxidation and hepatotoxicity [24]. As described in Rodrigo-Torres et al., acute CCl_4-treated mouse livers showed a mild induction of DR expansion. However, no duct-derived hepatocyte generation was observed based on dual staining of HNF4α/YFP (Fig. 3a).

1. Inject intraperitoneally a single dose of CCl_4 at 0.5 mL/kg, 10–25% diluted in corn oil.
2. Inject control mice with equivalent amount of corn oil.
3. Sacrifice 1 week after the injection.

3.2.4 Chronic Liver Damage Models with DDC Diet

DDC diet (Fig. 2d) induces liver damage and the expansion of DR and LPCs [25]. This model has been used to study the activation and proliferation of LPCs, but most especially it has been used as a model of primary sclerosing cholangitis and biliary fibrosis. In addition to DR, the phenotype observed in this model is characterized by fibrosis, inflammatory cells infiltration around periportal areas, cholestasis and expression of inflammatory cytokines.

In Rodrigo-Torres et al., after 4 weeks of DDC diet an important DR was observed, which was derived from HNF1β^+ cells. However, there was no contribution of HNF1β^+ cells to newly-formed hepatocytes. During recovery (2 weeks) with standard chow, DR was reduced, but once again, HNF1β^+ cells did not contribute to hepatocyte generation.

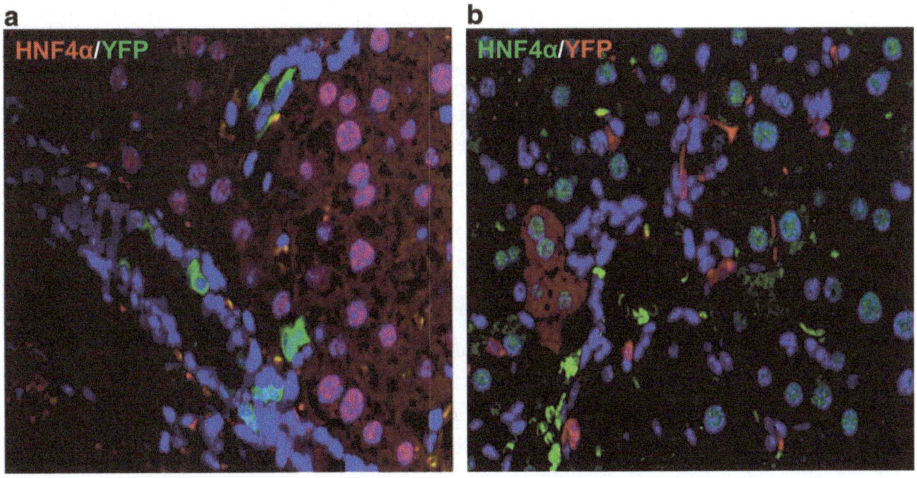

Fig. 3 Lineage tracing of HNF1β-expressing cells after liver injury. (**a**) In acute carbon tetrachloride (CCl_4) liver injury model, biliary epithelium cells (BECs) do not give rise to hepatocytes. Hepatocytes express nuclear hepatocyte nuclear factor 4 alpha (HNF4α) (red) and cells derived from hepatocyte nuclear factor 1-beta (HNF1β)+ BECs express yellow fluorescent protein (YFP) (green). (**b**) Image from choline-deficient ethionine-supplemented (CDE) diet liver injury model. Hepatocytes express HNF4α (green) and cells derived from HNF1β+ cells express YFP (red). A small number of hepatocytes express the reporter YFP indicating that they are derived from HNF1β+ cells

1. Feed mice ad libitum with a standard rodent chow diet containing 0.1% DDC for 4 weeks. To induce an injury-recovery model, feed mice for 4 weeks with DDC diet followed by 2 weeks of standard chow diet recovery (*see* **Note 4**).
2. Feed control animals with standard chow for the entire period.
3. Sacrifice mice when the protocol is completed.

3.2.5 Chronic Liver Damage Models with CDE Diet

CDE diet (Fig. 2e) is used as a model of expansion of DR cells. Unlike DDC diet, CDE diet does not induce biliary injury, but steatosis and a significant hepatocellular damage lead to failure of hepatocytes function [26]. CDE model also induces inflammatory cells proliferation and the expression of pro-inflammatory cytokines such as tumor necrosis factor alpha (TNF-α) and interleukin 1 beta (IL-1β).

In Rodrigo-Torres et al., after 3 weeks of CDE treatment, there was proliferation of DR cells derived from HNF1β^+ cells. Moreover, there was a small population of YFP$^+$ hepatocytes in the periportal area (0.22%). This percentage increased to 1.86% after 2 weeks of recovery (Fig. 3b).

1. Feed mice with CDE diet ad libitum for 3 weeks. To perform an injury-recovery model, feed the animals for 3 weeks with CDE diet followed by 2 weeks of standard diet (*see* **Note 4**).
2. Feed control animals with standard chow.
3. Sacrifice mice when the protocol is completed.

3.2.6 Chronic CCl_4 Administration

Long-term treatment with CCl_4 (Fig. 2f) induces hepatotoxicity leading to fibrosis, bile duct proliferation, and, eventually, hepatocellular carcinoma. Chronic CCl_4 administration is one of the most widely used models for fibrosis induction. CCl_4 treatment does not compromise hepatocyte replication. This model shows a limited expansion of DR.

1. Inject intraperitoneally CCl_4 at a dose of 0.5 mL/kg, twice a week for 8 weeks (*see* **Note 4**).
2. Inject control mice with equivalent amount of corn oil.
3. Sacrifice 4 days after the last injection.

3.3 Isolation of BEC-Derived Cells

Lineage tracing models allow isolating BEC-derived cells by flow cytometry cell sorting.

1. Anesthetize mice and cannulate inferior cava vein.
2. Perfuse livers with EGTA solution at a flow rate of 5 mL/min until liver gets pale.
3. Perfuse with collagenase solution containing Collagenase A 0.5 g/L at a flow rate of 5 mL/min for 11 min.
4. Mince digested liver in Petri dish with ice cold 1× HBSS.
5. Redigest with collagenase solution containing 0.5 g/L Collagenase A, 0.5 g/L pronase and 50 mg/L DNAse I.
6. Stir for 30 min at 37 °C.
7. Filter liver through 70-μm cell strainer.
8. Centrifuge at 100 × *g* for 2 min and keep supernatant.
9. Centrifuge supernatant at 300 × *g* for 5 min and keep the pellet.
10. Sort YFP+ cells by flow cytometry. Use cells from wild type (WT) mice to establish YFP^+ gate. Collect YFP^+ cells for further analysis.

3.4 Histological and Biochemical Analysis

1. Perform histochemical procedures, immunostaining, and colocalization studies using the abovementioned antibodies (*see* **Note 5**).
2. Evaluate apoptosis by TUNEL and caspases activity assays.
3. Analyze hepatic damage by measuring serum levels of ALT, AST, LDH and AP.

4 Notes

1. Cre expression may vary among different Cre mice. In addition, the efficiency of recombination could differ depending on the combination of CreER-expressing mouse and Cre-inducible Rosa26R reporter animal used for generating the lineage-tracing model. Therefore, it is important to evaluate the efficiency of recombination in each of the models generated.
2. GFP and YFP fluorescence cannot be distinguished by simply observing the liver sections under the microscope due to autofluorescence of the liver. Specific GFP/YFP signal should be detected using specific antibodies.
3. A high dose of tamoxifen is known to generate liver damage that could induce ectopic expression of the gene of interest and therefore of the reporter. It is important to optimize the amount of tamoxifen to ensure specific labeling. Intraperitoneal injection of tamoxifen is the most common administration method because the amount of administered compound can be better controlled. Doses ranging from 50 to 100 mg/kg are usually injected once per day for 5 consecutive days. In Rodrigo-Torres et al., mice were treated with 75 mg/kg, a widely used dose. Delivery by oral gavage is also possible. For oral administration the amount can range from 25 to 175 mg/kg for 5 consecutive days. Higher doses, ranging from 500 to 1000 mg/kg, can be used in shorter treatments: e.g., in Rodrigo-Torres et al., mice were treated on day 1 with 1000 mg/kg, on day 3 with 1000 mg/kg, and on day 5 with 500 mg/kg tamoxifen by oral gavage.
4. These protocols require daily monitoring of body weight. For DDC and CDE diet protocols, pellets could also be weighted to monitor food intake in each cage. Although a degree of weight loss is expected, rapid weight loss of 15–20% within a few days is a commonly accepted humane end point.
5. Immunostaining conditions should be optimized for every mouse model. Livers from mice fed with DDC diet contain large amounts of porphyrin crystals. These crystals are autofluorescent, leading to difficulties in immunofluorescent staining.

Acknowledgments

We want to thank Daniel Rodrigo-Torres for the advice during manuscript preparation and the experimental work that this chapter is based on. This work was supported by grants from the Fondo de Investigación Sanitaria Carlos III (FIS), co-financed by the Fondo Europeo de Desarrollo Regional (FEDER), Unión Europea, "Una manera de hacer Europa" (FIS PI14/00320, PI 17/00673 to PS-B). PS-B and BA-B are funded by the Instituto de Salud Carlos III, Miguel Servet (CPII16/00041) and PFIS, respectively, and co-financed by the Fondo Europeo de Desarrollo Europeo (FEDER), Unión Europea, "Una manera de hacer Europa."

References

1. Arias IM et al (2009) The liver: biology and pathobiology, 5th edn. John Wiley & Sons, Hoboken
2. Gouw ASH, Clouston AD, Theise ND (2011) Ductular reactions in human liver: diversity at the interface. Hepatology 54(5):1853–1863
3. Okabe M et al (2009) Potential hepatic stem cells reside in EpCAM+ cells of normal and injured mouse liver. Development 136 (11):1951–1960
4. Kholodenko IV, Yarygin KN (2017) Cellular mechanisms of liver regeneration and cell-based therapies of liver diseases. Biomed Res Int 2017:1–17
5. Lu W-Y et al (2015) Hepatic progenitor cells of biliary origin with liver repopulation capacity. Nat Cell Biol 17(8):971–983
6. Pu W et al (2016) Mfsd2a+ hepatocytes repopulate the liver during injury and regeneration. Nat Commun 7:13369
7. Schaub JR, Malato Y, Gormond C, Willenbring H (2014) Evidence against a stem cell origin of new hepatocytes in a common mouse model of chronic liver injury. Cell Rep 8 (4):933–939
8. Huch M et al (2013) In vitro expansion of single Lgr5+ liver stem cells induced by Wnt-driven regeneration. Nature 494(7436):247–250
9. Sancho-Bru P et al (2012) Liver progenitor cell markers correlate with liver damage and predict short-term mortality in patients with alcoholic hepatitis. Hepatology 55(6):1931–1941
10. Guy CD et al (2012) Hedgehog pathway activation parallels histologic severity of injury and fibrosis in human nonalcoholic fatty liver disease. Hepatology 55(6):1711–1721
11. Grompe M (2017) Fah knockout animals as models for therapeutic liver repopulation. Springer, Cham, pp 215–230
12. Carpentier R et al (2011) Embryonic ductal plate cells give rise to cholangiocytes, periportal hepatocytes, and adult liver progenitor cells. Gastroenterology 141(4):1432–1438.e4
13. Español-Suñer R et al (2012) Liver progenitor cells yield functional hepatocytes in response to chronic liver injury in mice. Gastroenterology 143(6):1564–1575.e7
14. Raven A et al (2017) Cholangiocytes act as facultative liver stem cells during impaired hepatocyte regeneration. Nature 547 (7663):350–354
15. Yanger K et al (2014) Adult hepatocytes are generated by self-duplication rather than stem cell differentiation. Cell Stem Cell 15 (3):340–349
16. Guest RV et al (2014) Cell lineage tracing reveals a biliary origin of intrahepatic cholangiocarcinoma. Cancer Res 74(4):1005–1010
17. Rodrigo-Torres D et al (2014) The biliary epithelium gives rise to liver progenitor cells. Hepatology 60(4):1367–1377
18. He L et al (2017) Enhancing the precision of genetic lineage tracing using dual recombinases. Nat Med 23(12):1488–1498
19. Solar M et al (2009) Pancreatic exocrine duct cells give rise to insulin-producing beta cells during embryogenesis but not after birth. Dev Cell 17(6):849–860
20. Kopp JL et al (2011) Sox9+ ductal cells are multipotent progenitors throughout development but do not produce new endocrine cells in the normal or injured adult pancreas. Development 138:653–665
21. Means AL, Xu Y, Zhao A, Ray KC, Gu G (2008) A CK19 CreERT knockin mouse line allows for conditional DNA recombination in epithelial cells in multiple endodermal organs. Genesis 46(6):318–323

22. Mitchell C, Willenbring H (2008) A reproducible and well-tolerated method for 2/3 partial hepatectomy in mice. Nat Protoc 3 (7):1167–1170
23. Kellendonk C, Tronche F, Casanova E, Anlag K, Opherk C, Schütz G (1999) Inducible site-specific recombination in the brain. J Mol Biol 285(1):175–182
24. Weber LWD, Boll M, Stampfl A (2003) Hepatotoxicity and mechanism of action of haloalkanes: carbon tetrachloride as a toxicological model. Crit Rev Toxicol 33 (2):105–136
25. Thompson MD, Awuah P, Singh S, Monga SPS (2010) Disparate cellular basis of improved liver repair in β-catenin-overexpressing mice after long-term exposure to 3,5-diethoxycarbonyl-1,4-dihydrocollidine. Am J Pathol 177(4):1812–1822
26. Itoh T, Miyajima A (2014) Liver regeneration by stem/progenitor cells. Hepatology 59 (4):1617–1626

Chapter 6

Specific Labeling and Lineage Tracing of Periportal Hepatocytes Using Two-Step Genetic Recombination

Nicola de Prisco, Eleanor Stout, and Joan Font-Burgada

Abstract

The liver is unmatched in regenerative capacity. However, when exhausted, the liver is predisposed to various diseases based on injury types and causal agents. Although hepatocytes have been proposed to be the main source of new hepatocytes during regeneration, the existence of specialized liver stem cells has been long debated. In mice, oval cells or ductal cells have been postulated as such stem/progenitor pool. Exhaustive works from different laboratories have shown that in genetically unmodified mice, oval cells, or by extension ductal cells, only contribute marginally in producing new hepatocytes during liver regeneration, thus indicating that hepatocytes are the main regenerative cell source. In this debated context, we identified a new population of periportal hepatocytes in the normal mouse liver. These cells we termed hybrid hepatocytes (HybHP) express low levels of the transcription factor Sox9. Using complementary lineage tracing tools, we demonstrated that HybHP regenerate the liver after chronic hepatocyte depleting injuries. Here, we describe the two-step genetic recombination method that allowed us to study HybHP's lineage in two established models of liver injury.

Key words Hybrid periportal hepatocytes, Dual recombinase, Lineage tracing, Liver regeneration

1 Introduction

Development of genetic lineage tracing techniques has shed light on previously unapproachable questions relating to complex tissue processes, stem cell dynamics, and cancer development. Liver biology has been broadly impacted by the implementation of lineage tracing in the quest to determine which cell populations are responsible for its staggering regeneration capacity. By tracing the lineage of oval cells, ductal cells, or hepatocytes in wild-type (WT) mice, it has become evident that the main source of new hepatocytes in a large variety of mouse models of liver regeneration is hepatocytes themselves [1–6]. The negligible contribution found from oval/ductal cells has challenged the notion of a dedicated liver progenitor pool. However, in genetically modified mice that lack β1-integrin in hepatocytes or that overexpress p21 in hepatocytes, bile duct

Naoki Tanimizu (ed.), *Hepatic Stem Cells: Methods and Protocols*, Methods in Molecular Biology, vol. 1905, https://doi.org/10.1007/978-1-4939-8961-4_6, © Springer Science+Business Media, LLC, part of Springer Nature 2019

cells produce a significant proportion of new hepatocytes after injury cessation by outcompeting original hepatocytes with blocked proliferation potential [7]. Recently, we identified a population of periportal hepatocytes in the normal mouse liver which express low levels of Sox9, a highly expressed transcription factor in bile duct cells [8]. These so-called hybrid hepatocytes (HybHP) are quiescent during homeostasis. However, we found that in different mouse models of hepatocyte injury, HybHP are the cells that produce the majority of the new hepatocytes that repopulate the liver. In addition, during cholestatic injury, few metaplastic HybHP fully transdifferentiate to ductal phenotype. Since Sox9-Cre^{Ert2} was targeting both bile duct cells and HybHP and there was no other tool to specifically label the latter, we developed a new protocol based on the sequential utilization of two independent recombinases to clonally label HybHP.

For this purpose, we crossed Sox9-Cre^{Ert2} with NZG mice which contain both a loxP-flanked STOP PGK neo cassette upstream of a nuclear targeted LacZ marker and a downstream GFP marker, whose expression is prevented by the Frt-flanked LacZ cassette itself (Fig. 1). The Sox9-Cre^{Ert2} transgene expressed tamoxifen-inducible Cre in both ductal cells and HybHP. Thus, after administration of tamoxifen to Sox9-Cre^{Ert2};NZG heterozygous mice, the floxed STOP cassette was recombined in ductal cells and HybHP with concomitant expression of nuclear LacZ with no cells expressing eGFP (Fig. 2a–c). The sequential used adeno-associated virus (AAV) TBG-FLPo which only transduce and express the recombinase in hepatocytes. Transduction of tamoxifen-treated Sox9-Cre^{Ert2};NZG heterozygous mice with AAV-TBG-FLPo resulted in the excision of LacZ cassette in all hepatocytes and only allowed eGFP expression by HybHP, which had previously recombined the STOP cassette (Figs. 1 and 3a). Staining with different markers for ductal-hepatocyte lineages confirmed the specificity of HybHP GFP labeling (Fig. 3b, c).

Once clonal HybHP were GFP labeled, these mice were used to perform lineage tracing in different models of liver injury, confirming that HybHP play a significant role in repopulating the liver after chronic hepatocyte damage. This two-step genetic recombination protocol can be used to label any population of cells lacking a specific marker if these cells can be defined by the unique intersection of two independent markers that otherwise are shared with other cell types.

2 Materials

2.1 Mice Strains

1. Sox9-Cre^{Ert2}: This strain was described in Kopp et al., 2011 [9]; Sox9-Cre^{Ert2} mice express Cre^{Ert2} recombinase in all Sox9-expressing cells, which in the liver are both hybrid periportal hepatocytes and bile duct cells.

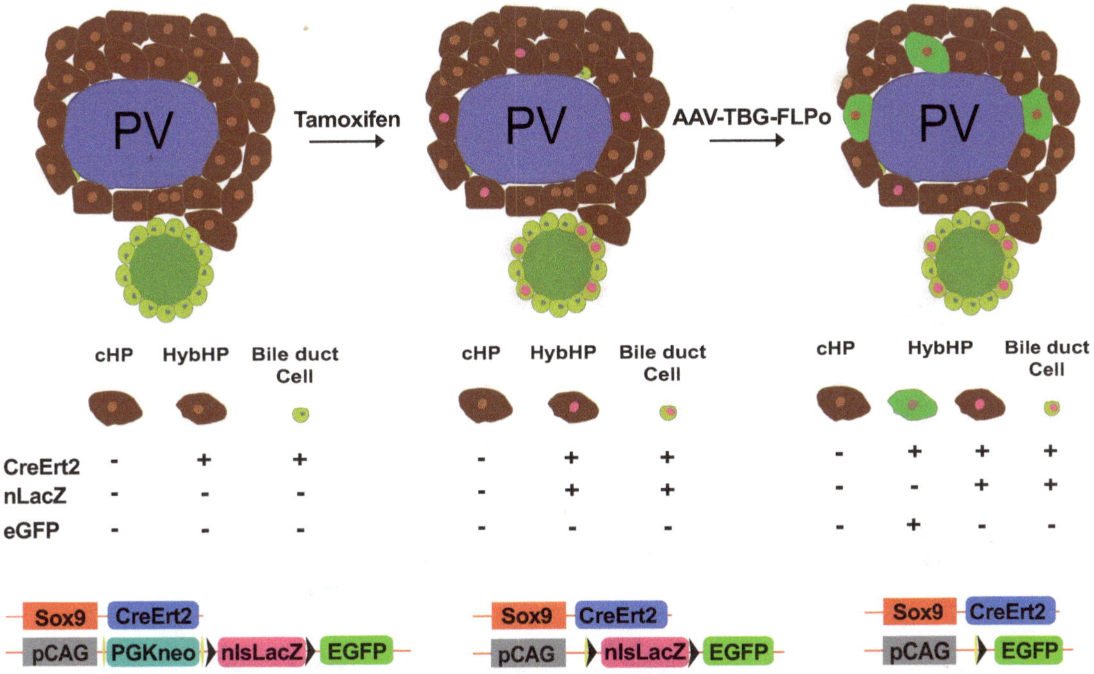

Fig. 1 Specific labeling of HybHP using NZG reporter. Schematic outlining of the two-step genetic recombination using NZG reporter. Sox9-Cre^{Ert2};NZG mice express inducible Cre in ductal cells and HybHP (left panel). Tamoxifen treatment induces Cre-mediated recombination of PGK neo cassette with the subsequent expression of nuclear LacZ in ductal cells and HybHP (pink nuclei, middle panel). When Sox9-Cre^{Ert2};NZG mice previously treated with tamoxifen are injected with adeno- or AAV-FLPo, which is expressed only in hepatocytes, the recombination of nuclear LacZ and the subsequent expression of eGFP, only in HybHP, occur (right panel)

2. NZG: Jax 012429 [10] contains loxP-flanked STOP PGK neo cassette that prevents the expression of a nuclear-targeted LacZ marker. Furthermore, Frt sites flanking LacZ prevent in turn the downstream expression of eGFP (Fig. 1).

2.2 Models of Liver Injury

1. Sox9-Cre Ert2;NZG previously treated with tamoxifen and injected with AAV expressing FLPo were treated with carbon tetrachloride (CCl_4) to induce liver injury and determine the clonal behavior of HybHP. Mice are treated with a single dose (acute) of CCl_4 or with 6 and 12 injections of CCl_4.
2. To induce cholestatic liver injury and see if HybHP can give rise to true ductal cells, Sox9-Cre ^{Ert2};NZG previously treated with tamoxifen and with the AAV were fed with diethoxycarbonyl-1,4-dihydrocollidine (DDC) diet for 6 weeks.

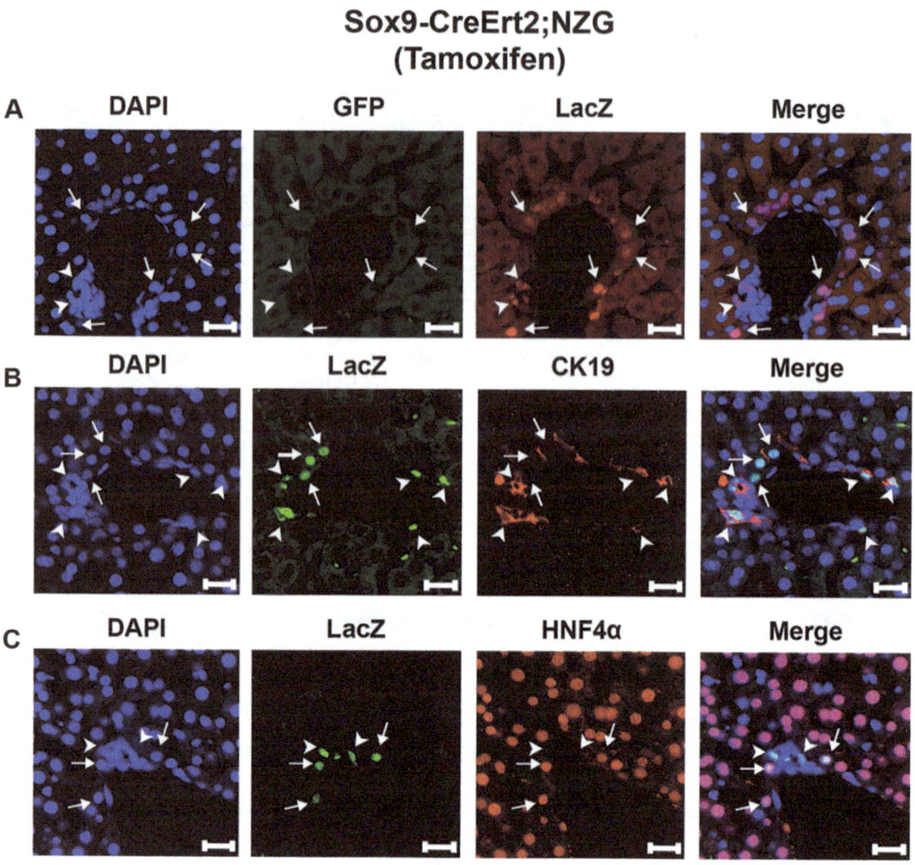

Fig. 2 First recombination step. (**a**) Sox9-Cre^{Ert2};NZG mice (4–6 weeks old) were injected with 100 mg/kg of tamoxifen. Ductal cells and HybHP were positive for nuclear LacZ (arrowheads and arrows, respectively). (**b**) LacZ^{+} ductal cells were CK19^{+} (arrowheads) and HybHP LacZ^{+} were CK19^{-} (arrows). (**c**) LacZ^{+} HybHP were positive for HNF4α (arrows), while LacZ^{+} ductal cells were HNF4α^{-}. Scale bar 20 μm

2.3 Administration of Adenoviral and Adeno-Associated Viral Vectors Expressing FLPo Recombinase

1. Recombinant adenovirus expressing FLP recombinase (Ad-CMV-FLPo, Vector Biolabs #1775).
2. Recombinant AAV-DJ-TBG-FLPo (Vector Biolabs #1728).
3. 2 L water bath set at 48 °C.
4. Mouse restrainer.
5. 1 mL syringe.
6. 27G needle.
7. 70% ethanol.

2.4 Mouse Weighing and Tamoxifen Preparation

1. Scale.
2. Plastic container or plastic beaker.
3. Tamoxifen.
4. Corn oil.

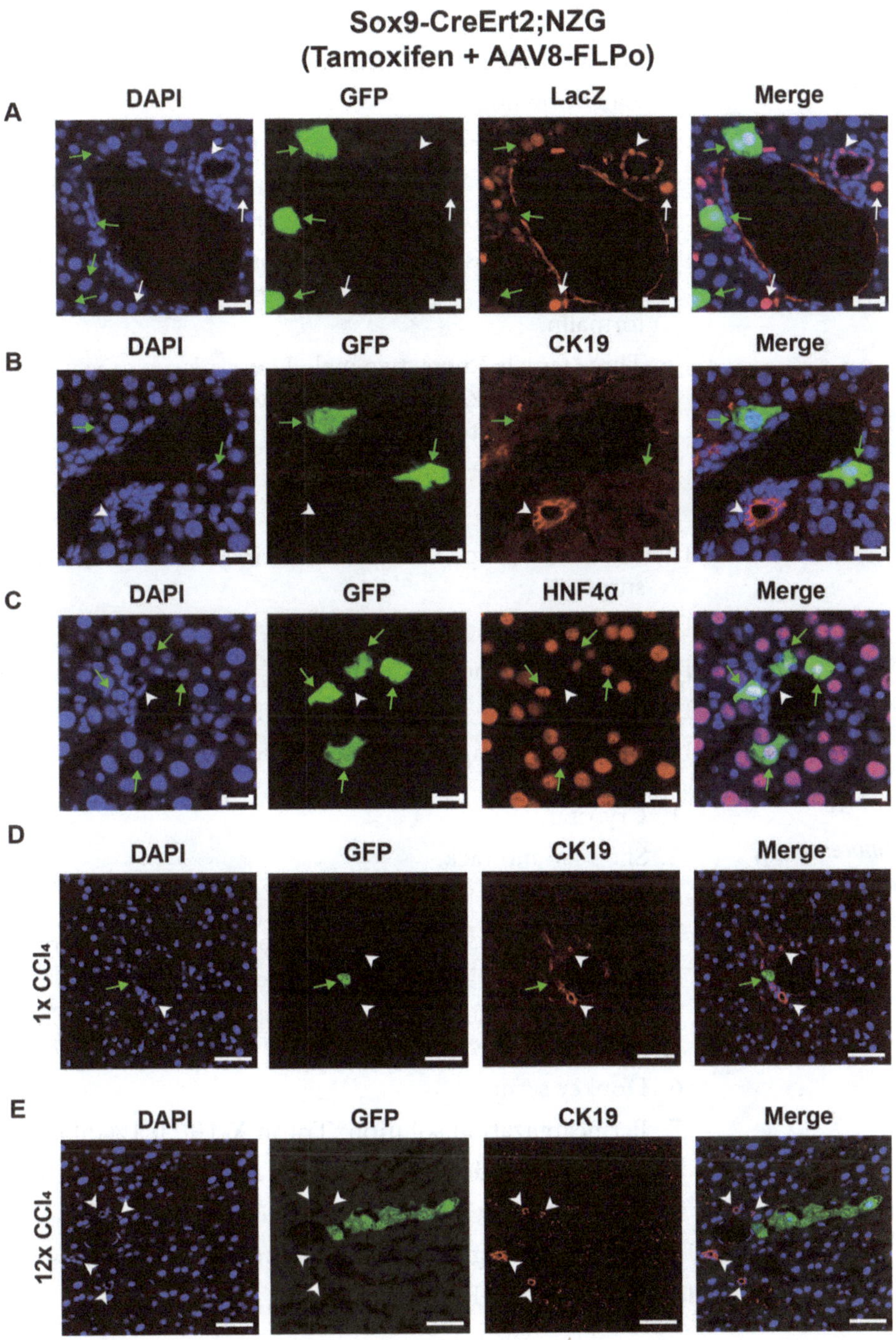

Fig. 3 Second recombination step. (**a**) Sox9-Cre^{Ert2};NZG mice were first treated with 100 mg/kg of tamoxifen and 10 days later were given 20 mg/kg tamoxifen. Two weeks later, mice were given 10^9 adeno- or 5×10^{11} AAV-FLPo viral particles, respectively. After 10 days liver sections were obtained, stained, and imaged. Only some HybHP were GFP$^+$ (green arrows), while some HybHP were not recombined by FLPo recombinase, remaining nuclear LacZ$^+$ and GFP$^-$ (white arrow). Ductal cells were GFP$^-$ and nuclear LacZ$^+$ (arrowhead). After 10 days some GFP$^+$ HybHP still had residual nuclear LacZ signal. (**b**) All GFP$^+$ HybHP were CK19$^-$ (green

2.5 Mouse Heart Perfusion

1. DPBS.
2. Heparin: 10 KU/mL heparin in DPBS (*see* **Note 1**).
3. Zinc formalin.
4. Form-Zero neutralizing powder (Fisher Scientific).
5. Tris–HCl pH 9.5/DTT solution: 100 mM Tris pH 9.5, 10 mM DTT solution (Teknova).
6. Syringe pumps (NewEra Pumps #NE300): two pumps are required, one for DPBS-heparin and the other for zinc formalin.
7. Three female Luers, two male Luers, plastic tube with the inner diameter of 1.6 mm, and three-way stopcocks (Biorad, #732-8103) (*see* **Note 2**).
8. Formalin absorbent pads (Leica, large pads).
9. Isoflurane.
10. Scissors, clamp, forceps, and circle tip forceps: for mouse surgery.
11. Surgery tray.
12. Sucrose: 15 and 30% in PBS.
13. Cryomold: 25 mm × 20 mm × 5 mm.
14. Optimal cutting temperature (OCT) compound.

2.6 Immunofluorescence, Antibodies, and Staining Reagents

1. Cryostat.
2. Slide Jar and rack.
3. Staining tray.
4. Citrate buffer: Dilute 10× citrate buffer pH 6.0 (Scytek #CPL500) with MilliQ water.
5. Phosphate saline buffer (PBS): 137 mM NaCl, 10 mM Na_2HPO_4, 1.8 mM KH2PO4, 2.7 mM KCl, pH 7.3–7.5.
6. Donkey serum.
7. Permeabilization solution: Triton X-100 0.1% diluted in PBS.
8. Blocking solution: Tween 0.1%, donkey serum 2% in PBS.
9. Wash solution: Tween 0.1% in PBS.
10. Mounting reagent: Mowiol or an equivalent reagent.
11. Goat anti-HNF-4α (Santa Cruz, sc-6556), 1:50.

Fig. 3 (continued) arrows). (**c**) All GFP^+ HybHP were $HNF4\alpha^+$ (green arrows). Sox9-Cre^{Ert2};NZG were given tamoxifen followed with AAV-FLPo and treated with 1 or 12 (chronic) CCl_4 doses. Liver was excised and analyzed as above. Single GFP^+ HybHP were found after only one CCl_4 dose (**d**), while multicellular GFP^+ hepatocytes clones expanding from the portal tract (PT) were observed after 12 CCl_4 doses (**e**). Bracketed and open scale bars: 20 and 50 μm, respectively

12. Goat anti-CK19 (Santa Cruz, sc-33,111), 1:100.
13. Mouse anti-LacZ (Promega, Z3781), 1:50.
14. Chicken anti-eGFP (Abcam, Ab13970), 1:800.
15. 1 mg/mL DAPI stock solution.

3 Methods

Sox9 Cre^{Ert2} is not viable in homozygosity. This stock is maintained by breeding heterozygous males with C57/BL6 females.

NZG colony is maintained by breeding homozygous mice.

For experiments, Sox9-Cre^{Ert2} males are bred with homozygous NZG females producing offspring 100% heterozygous for the NZG reporter and 50% for Sox9-Cre^{Ert2}. Experiments are performed with heterozygous Sox9-CRE^{Ert2};NZG mice.

3.1 Tamoxifen Preparation and Administration

Injection of Sox9-Cre^{Ert2};NZG mice at postnatal day 10 intraperitoneally at the dose of 100 mg/kg body weight (*see* **Note 3**).

1. Under a chemical fume hood weight 20 mg of tamoxifen.
2. Dissolve the tamoxifen in 1 mL of sterile corn oil obtaining a solution of 20 mg/mL (*see* **Note 4**).
3. Using a P1000 pipette, dispense 500 μL of tamoxifen directly into the barrel of the syringe and carefully and slowly pull the plunger back. Invert the syringe and expel air bubbles.
4. Weigh the mouse to be injected, and inject intraperitoneally the corresponding volume of the solution to administrate 100 mg/Kg tamoxifen.
5. Maintain the injected mice on absorbent paper for 72 h post injection then change to a clean cage (*see* **Note 5**).

3.2 Administration of Adenovirus or Adeno-Associated Virus via Tail Vein Injection

Transduction of Sox9-CRE^{Ert2};NZG mice with Ad-/AVV-DJ-FLPo after a minimum of 2 weeks from tamoxifen administration to wash out tamoxifen. After a minimum wash-out period of 2 weeks from virus administration, the mice are ready to lineage trace HybHP clonally with the model of liver injury of choice.

1. Before the intravenous injection (i.v.), dilute the virus in PBS. Use 5×10^{11} genome copies/mouse for AAV or 1×10^{9} plaque-forming units/mouse of adenovirus (*see* **Note 6**).
2. Prepare and set a water bath at 48 °C.
3. Load the syringe with 200 μL of the solution containing the adenovirus or the AAV.
4. Place the mouse in the mouse restrainer.

5. Hold the tail in the warm water for 45 s to 1 min to dilate the tail vein.
6. Inject in one of the two lateral tail veins.

3.3 Intracardiac Perfusion and Staining

Carry out all the procedure under chemical hood.

Day 1 (*see* **Note 7**)

1. Prepare a 50 mL syringe with 45 mL of DPBS-heparin, and load the syringe into the pump.
2. Prepare a 50 mL syringe with 45 mL of zinc formalin and load in the second syringe pump.
3. Prepare the surgery tray with the formalin absorbent pad.
4. Prepare 50 mL conical tube with 30 mL of zinc formalin.
5. Anesthetize the mouse with isoflurane using anesthesia chamber.
6. Pin the mouse on the formalin absorbing pad in the surgery tray and spray the abdominal area with 70% ethanol.
7. Using the circle tip forceps and scissors, cut the skin up from the stomach along the side around the heart, then carefully cut the peritoneal membrane.
8. Carefully cut the diaphragm and the rib cage both sides of the sternum. Clamp the sternum, and roll it to expose the heart.
9. Insert the pump needle into the left ventricle.
10. Once the needle is placed in the left ventricle, cut the right auricle with scissors, and immediately start pumping DPBS-heparin solution at 5 mL/min.
11. Pump 15–20 mL DPBS-heparin, and once the liver appears clear, turn off the pump, and switch the valve to zinc formalin.
12. Perfuse the mouse with 20 mL of zinc formalin (*see* **Note 8**), then stop the pump, and remove the needle.
13. Excise the liver, and put it into a 50 mL conical tube with 30 mL zinc formalin, and gently rock it overnight at 4 °C.
14. Repeat the procedure for all mice by washing the tubing with DPBS for one full flush followed by one full flush of DPBS-heparin solution.

Day 2

15. Discard the zinc formalin, and rinse twice the liver filling the 50 mL conical with PBS, mix, and discard (*see* **Note 9**).
16. Wash with PBS for 15 min at room temperature (RT), rocking gently and repeat this wash twice.
17. Rinse quickly with MilliQ water.

18. Using a 10 cm plate as tray, carefully separate each lobe, cut lobes in half, and put them into a new 15 mL tube.
19. Add 10 mL of Tris–HCl pH 9.5/DTT solution.
20. Rock gently at RT for 2 h.
21. After 2 h discard the Tris–HCl pH 9.5/DTT solution.
22. Do two quick washes with PBS.
23. Fill the tube with 10 mL of 15% sucrose.
24. Rock gently at 4 °C overnight.

Day 3

25. Discard 15% sucrose.
26. Add 10 mL of 30% sucrose, rocking gently at 4 °C until the next day.

Day 4

27. Prepare a box with dry ice, OCT compound and standard size cryomolds.
28. Pour the 30% sucrose solution from a liver sample, then pour the liver pieces onto a 10 cm plate.
29. Pour a small amount of OCT compound in two standard size cryomolds, enough to cover the surface.
30. Stack a Kimwipe on a paper towel, place the liver pieces on the Kimwipe, and dry off with another Kimwipe on top, gently pressing to collect all the sucrose solution (*see* **Note 10**).
31. Place each half of liver lobes into separated cryomolds, with the convex side of each piece facing down, and cover completely with OCT compound.
32. Place the cryomold onto dry ice until the OCT is completely frozen, making sure the mold is completely flat so the OCT compound does not spill over.
33. Store samples in −80 °C.
34. Cut liver sections of 5 μm using a cryostat.

3.4 Immunofluorescence

Day 1

1. Set the water bath to 96 °C.
2. Pour the 1× citrate buffer into staining jar. Once the water bath reaches the temperature, place the staining jar with citrate buffer inside.
3. Place sample slides in a slide rack, and do three PBS washes of 5 min each using an additional staining jar.
4. Perform two quick washes with MilliQ water, then tap the water off onto a paper towel, and place the slides into citrate buffer at 96 °C for 1 h.

5. After 1 h, take the staining jar out of the water bath and let cool at room temperature for 45 min.
6. Do two quick washes with PBS.
7. Prepare 100 mL of 0.1% Triton X-100 in PBS, and incubate the slides for 20 min in a staining jar.
8. After 20 min, do five quick washes with PBS (*see* **Note 11**).
9. Process one slide a time, drawing a circle around the tissue section with Dako pen and then incubate with 150–200 μL of blocking buffer, carefully avoiding the tissue to dry for 30 min in a staining tray.
10. Prepare appropriate antibody solution by diluting antibodies in blocking buffer (*see* **Note 12**).
11. After 30 min, pour the blocking buffer, and carefully add the antibodies to each slide sample.
12. Incubate at 4 °C overnight.

Day 2

13. Wash slides three times for 5 min each with wash solution using a staining jar.
14. Incubate each slide with the corresponding secondary antibody diluted in blocking buffer for 2 h at RT.
15. Wash three times for 5 min with wash solution using a staining jar.
16. Wash two times for 5 min with PBS.

Optional if reduction of liver autofluorescence is desired; otherwise follow directly to **step 20**:

17. Wash the slides further with two quick washes with MilliQ water and one quick wash with 70% ethanol.
18. Incubate slides with 0.1% Sudan Black in 70% ethanol for 20–30 min.
19. After this incubation, thoroughly wash slides with PBS 0.02% Tween, and using a squeeze bottle wash applying directly the jet to the tissue being careful not to wipe it out.
20. Incubate each slide with DAPI diluted 1:1000 in PBS (1:100 in PBS-0.02% Tween if Sudan Black is used) for 10 min for nuclear staining.
21. Wash quickly the slides using a staining jar with MilliQ water. If Sudan Black has been used, wash the slides three times for 5 min with PBS 0.02% Tween, followed by two quick washes with MilliQ water.
22. Tap each slide dry on a paper towel, and then tip the slides up in a drawer to finish drying and be protected from the light.
23. Once dried, mount slides using 10 μL of Mowiol.

4 Notes

1. Add 10 mL of nuclease-free water in 100KU heparin bottle (Sigma), mix thoroughly, transfer into a 15 mL conical tube, and store at 4 °C. Dilute heparin (1:1000) in DPBS to obtain the working DPBS-heparin solution.
2. Connect the two syringe pumps (one with DPBS-heparin solution and the other with zinc formalin) to 1.6 mm plastic tube using female luers. Connect the two tubes with the three-way stopcock using the male luers, and finally connect the three-way stopcock with a 1.6 mm tube with a 26½ needle.
3. Mice can be injected at any age, but in older mice, recombination becomes less efficient. Consecutive tamoxifen injection to maximize the number of Sox9-expressing cells that recombine can be successfully performed.
4. To dissolve properly the tamoxifen, prewarm a thermomixer at 55 °C, and put the tube containing corn oil and tamoxifen for 30 min with mixing (speed 800–900 rpm). Remaining crystals of tamoxifen should dissolve in 30 min.
5. Dispose the bedding into a yellow Chemotherapy Waste bag.
6. Even though we have successfully labeled HybHP with both adenovirus and AAV, we have observed that adenovirus is associated with a clear immune reaction, not detected using AAV, that could be potentially affecting downstream experiments.
7. All volumes per one mouse.
8. If the perfusion is working properly, the mouse tail will get stiff, and at the end of the perfusion, the liver should be stiff to the touch as well.
9. Collect the zinc formalin into a waste bottle, and neutralize it by adding Form-Zero formaldehyde powder neutralizer. Add the entire content of the bottle (500 cc) for each 4 L of formalin waste, mix the solution, and let it stand for 20–25 min to allow the reaction to complete. Dispose the neutralized formalin into the sanitary sewer.
10. Residual sucrose from the tissue must be removed before embedding. Sucrose will prevent OCT to contact the tissue affecting the quality of the histological sectioning.
11. At this point, the slides can be stored in PBS at 4 °C for few days.
12. Prepare the proper antibody dilution in a final minimum volume of 50 μL of blocking buffer per slide.

Acknowledgments

This work has been supported by CIRM Training Grant II (TG2-01154) and NIH K99/R00CA191152 grant.

References

1. Rodrigo-Torres D, Affo S, Coll M, Morales-Ibanez O, Millan C, Blaya D, Alvarez-Guaita A, Rentero C, Lozano JJ, Maestro MA, Solar M, Arroyo V, Caballeria J, van Grunsven LA, Enrich C, Gines P, Bataller R, Sancho-Bru P (2014) The biliary epithelium gives rise to liver progenitor cells. Hepatology 60 (4):1367–1377. https://doi.org/10.1002/hep.27078
2. Schaub JR, Malato Y, Gormond C, Willenbring H (2014) Evidence against a stem cell origin of new hepatocytes in a common mouse model of chronic liver injury. Cell Rep 8 (4):933–939. https://doi.org/10.1016/j.celrep.2014.07.003
3. Tarlow BD, Finegold MJ, Grompe M (2014) Clonal tracing of Sox9+ liver progenitors in mouse oval cell injury. Hepatology 60 (1):278–289. https://doi.org/10.1002/hep.27084
4. Yanger K, Knigin D, Zong Y, Maggs L, Gu G, Akiyama H, Pikarsky E, Stanger BZ (2014) Adult hepatocytes are generated by self-duplication rather than stem cell differentiation. Cell Stem Cell 15(3):340–349. https://doi.org/10.1016/j.stem.2014.06.003
5. Espanol-Suner R, Lemaigre FP, Leclercq IA (2013) Reply: to PMID 22922013. Gastroenterology 145(1):255–256. https://doi.org/10.1053/j.gastro.2013.05.037
6. Malato Y, Naqvi S, Schurmann N, Ng R, Wang B, Zape J, Kay MA, Grimm D, Willenbring H (2011) Fate tracing of mature hepatocytes in mouse liver homeostasis and regeneration. J Clin Invest 121 (12):4850–4860. https://doi.org/10.1172/JCI59261
7. Raven A, Lu WY, Man TY, Ferreira-Gonzalez S, O'Duibhir E, Dwyer BJ, Thomson JP, Meehan RR, Bogorad R, Koteliansky V, Kotelevtsev Y, Ffrench-Constant C, Boulter L, Forbes SJ (2017) Cholangiocytes act as facultative liver stem cells during impaired hepatocyte regeneration. Nature 547(7663):350–354. https://doi.org/10.1038/nature23015
8. Font-Burgada J, Shalapour S, Ramaswamy S, Hsueh B, Rossell D, Umemura A, Taniguchi K, Nakagawa H, Valasek MA, Ye L, Kopp JL, Sander M, Carter H, Deisseroth K, Verma IM, Karin M (2015) Hybrid periportal hepatocytes regenerate the injured liver without giving rise to cancer. Cell 162(4):766–779. https://doi.org/10.1016/j.cell.2015.07.026
9. Kopp JL, Dubois CL, Schaffer AE, Hao E, Shih HP, Seymour PA, Ma J, Sander M (2011) Sox9 + ductal cells are multipotent progenitors throughout development but do not produce new endocrine cells in the normal or injured adult pancreas. Development 138 (4):653–665. https://doi.org/10.1242/dev.056499
10. Yamamoto M, Shook NA, Kanisicak O, Yamamoto S, Wosczyna MN, Camp JR, Goldhamer DJ (2009) A multifunctional reporter mouse line for Cre- and FLP-dependent lineage analysis. Genesis 47(2):107–114. https://doi.org/10.1002/dvg.20474

Chapter 7

Analysis for the Heterogeneity of Liver Progenitor Cells

Kenji Kamimoto

Abstract

Recent technological advances have revealed the heterogeneity of cells and tissues. Existence of heterogeneity in hepatic progenitor cells is becoming apparent by various experimental approaches, and here we describe a series of techniques to investigate the proliferative heterogeneity of these cells. We have developed a new technique by combining genetic lineage tracking and three-dimensional imaging methods. The data obtained can be used in statistical analysis to quantitatively investigate the mechanisms underlying the heterogeneity of hepatic progenitor cells.

Key words Heterogeneity, Cell proliferation, Three-dimensional imaging, Microscopy, Genealogy tracking, Quantitative analysis

1 Introduction

Recent technological advances in dissecting biological events at the single-cell level have revealed the existence of heterogeneity in biological aspects such as gene expression, cellular function, and proliferative capacity [1, 2].

Hepatic progenitor cells, which are characterized to have bipotential differentiation and high proliferative capacities [3], have also been hypothesized to be heterogeneous [4]. Its proliferative heterogeneity has been reported in many studies using in vitro colony formation assays [5–7]. Proliferative capacity is examined by measuring the size of a colony derived from a single hepatic progenitor cell [8]. Cell culture is initiated at a very low density so that their progenies appear as colonies, and this feature allows retrospective identification of the original proliferative and differentiation capacities of each progenitor cell. Therefore, colony formation assays provide useful information on hepatic progenitor cells. However, colony behavior revealed by this method does not necessarily reflect its in vivo physiological capacity, since the assay conditions may not recapitulate the situation around progenitor cells within the tissue of origin. Here we describe a series of methods for quantitatively

Naoki Tanimizu (ed.), *Hepatic Stem Cells: Methods and Protocols*, Methods in Molecular Biology, vol. 1905, https://doi.org/10.1007/978-1-4939-8961-4_7, © Springer Science+Business Media, LLC, part of Springer Nature 2019

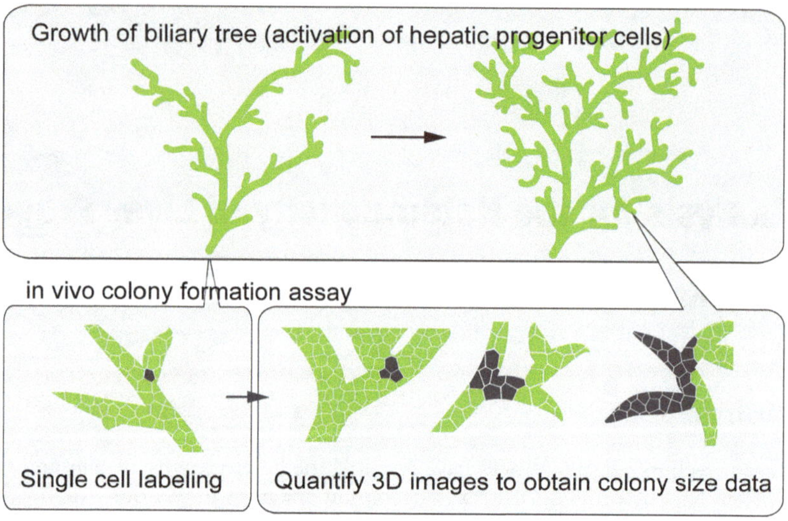

Fig. 1 Schematic illustration of the proposed method

evaluating the proliferative capacity of hepatic progenitor cells in situ.

In vivo investigation of proliferative capacity of hepatic progenitor cells is achieved by combining clonal genetic fate tracing with three-dimensional (3D) microscopy [9] (Fig. 1). Hepatic progenitor cells can be labeled in vivo and traced using a genetic lineage tracing system [10]. In our proposed method, we utilize Cre-loxP system, which is a frequently used genetic lineage tracing system in mouse studies. Thus, cell labeling can be performed at the desired frequency. We do not need to isolate progenitor cells for the analysis of cell proliferation because we can label and trace target cells in vivo. Besides, inducible Cre-loxP system can be tunable. Thus we can induce cell labeling at the desired frequency. Importantly, the fate of single cells can be traced by labeling at a frequency low enough to distinguish each labeled cell. This sparse permanent label makes it possible to track the fate of individual hepatic progenitor cells under physiological conditions.

Tracing sparsely labeled cells has been applied for the study of epidermal tissue [11–13], which shows a structure composed of simple stratified layers. In such a structure, the colony derived from a single stem cell spreads in 2D, and therefore, its size can be measured easily. In contrast, the niche of hepatic progenitor cells is the intrahepatic biliary system, which is the 3D tissue structure within the liver tissue (Fig. 1). In order to apply the in vivo quantitative clonal cell tracing system to the investigation of hepatic progenitor cells, we established the high-resolution 3-D imaging technique visualizing the intrahepatic biliary system at single-cell level [14].

Data is acquired from numerous colonies, each derived from a single hepatic progenitor cell. By analyzing the data, the mechanisms underlying proliferative heterogeneity can be investigated [9].

2 Materials

2.1 Mouse Experiment

1. Mouse strain: Prom1-CreERT2. This mouse strain has the CreERT2 gene at the Prom1 locus [15] and may be purchased from the Jackson Laboratory (Stock No: 017743).
2. Mouse strain: R26R-tdTomato. This mouse strain has a loxP-flanked transcriptional stop fragment inserted upstream of the red fluorescent protein (tdTomato) gene [16]. The strain is used in combination with the Cre-expressing mouse strain for lineage tracing experiments and may be purchased from the Jackson Laboratory (Stock No: 007909).
3. Mouse strain: Prom1-CreERT2;R26R-tdTomato. Cross Prom1-CreERT2 with R26R-tdTomato mouse strains to obtain a Prom1-CreERT2;R26R-tdTomato mouse (*see* **Note 1**).
4. Tamoxifen solution: Dissolve 10 mg of tamoxifen in 10 mL corn oil to obtain a 1 mg/mL solution. It takes several hours to completely dissolve tamoxifen powder in corn oil. Place the mixture on an agitator at 37 °C, and incubate it until the powder is dissolved.
5. Gavage needle.
6. Thioacetamide (TAA) solution: Dissolve TAA in water at 300 mg/L. Prepare this solution just before administration (*see* **Note 2**).
7. 26G syringe needle.
8. 10 mL syringe.
9. CO_2 cylinder.

2.2 Tissue Processing

1. 4% paraformaldehyde (PFA) in phosphate-buffered saline (PBS): Dissolve PFA in PBS and store at 4 °C.
2. 20% sucrose in PBS: Dissolve sucrose in PBS and store at 4 °C.
3. OCT compound.
4. Liquid nitrogen.
5. Cryostat.
6. 5 mL tube.
7. Permeabilization buffer: 3% bovine serum, 0.02% sodium azide, and 0.2% Triton X-100 in PBS. Store at 4 °C.

8. Nucleus staining solution: Dilute 5 mM SYTOX Green nucleic acid stain in permeabilization buffer at a 1:100000 ratio (*see* **Note 3**). Prepare the solution just before use.
9. SeeDB solution [14]: 80.2% D-fructose (wt/wt) and 0.5% 1-thioglycerol in water. Prepare just before use (*see* **Note 4**).
10. Sample storage solution: 3% bovine serum, 0.02% sodium azide, and 20% sucrose in PBS.
11. Confocal microscopy: Use a laser confocal microscope equipped with silicone or oil immersion lens (e.g., microscope, FV1200; and silicone immersion lens, UPLSAPO30XS) (*see* **Note 5**).

3 Methods

3.1 Mouse Experiment

1. Obtain Prom1-CreERT2;R26R-tdTomato mice. Six- to eight week-old animals are desirable (*see* **Note 6**).
2. Administer 250 μL of tamoxifen solution per 20 g body weight to the mouse by oral gavage (*see* **Notes 7** and **8**) with the 1.0 mL syringe and gavage needle.
3. Start liver injury 2 weeks after tamoxifen administration. For a TAA liver injury model, use the TAA solution, which can be administrated as drinking water (*see* **Note 9**).
4. Change TAA solution every week to keep it fresh.

3.2 Tissue Fixation

The mouse liver should be harvested several weeks after TAA administration. We recommend analyzing the liver after 1–8 weeks. The experimental period should be selected according to the experimental design. Detailed information can be inferred from analyzing a range of injury periods rather than a fixed time point.

1. Place mouse in chamber for CO_2 euthanasia.
2. Flow high concentrations of CO_2 until vital signs are lost.
3. Immediately following euthanasia, set the mouse for surgery to start PFA perfusion (*see* **Note 10**).
4. Remove the skin and muscle layers of the abdomen using surgical scissors to expose the liver.
5. Cut the vena cava to let the blood flow out.
6. Insert the syringe needle into the portal vein, and inject 10 mL of PBS gently until blood flows out from the liver. Next, inject 10 mL of 2% PFA and then 10 mL of 4% PFA. Make sure the liver color changes into light brown and the liver becomes hard (*see* **Note 11**).
7. Remove the liver from the mouse body.

8. Remove the gallbladder and the extrahepatic bile duct using surgical scissors.
9. Separate the hepatic lobules using a surgical scissor or knife.

3.3 Tissue Processing

1. Cut the liver into ~1 cm thick slices with a razor blade.
2. Place the slices in a 5 mL tube filled with 4% PFA, and incubate at 4 °C overnight. Remove the PFA and wash the liver twice with PBS. Incubate the liver slices in 20% sucrose at 4 °C overnight.
3. Embed the fixed liver slices in the OCT compound. Freeze the sample with liquid nitrogen. The frozen sample can be stored at −80 °C.
4. Place the frozen liver into the sample holder of the cryostat.
5. Slowly warm the frozen liver to melt the surface of the sample by touching with the hand. Immediately after thawing the surface of the sample, cut the frozen liver into 200–300 μm thick sections using the cryostat (*see* **Note 12**).
6. Using forceps, place the samples into PBS in a 50 mL tube. At least 30 liver slices should be obtained from a frozen liver block with 1 cm in thickness (*see* **Note 13**).
7. Wash the sliced samples with PBS three times.
8. Wrap the tube in aluminum foil sheet, or incubate in the dark to avoid photobleaching of fluorescent dye and the fluorescent protein.
9. Place the liver samples into a 5 mL tube filled with 2 mL of permeabilization buffer. Incubate the samples for 30 min on a rocking device (*see* **Note 14**).
10. Aspirate the permeabilization buffer from the tube, and add 2 mL of nucleus staining solution. Incubate the tube overnight on a rocking device (*see* **Note 15**).
11. Wash samples with PBS three times.
12. Remove the PBS completely. Mix samples with 2 mL of SeeDB in a new 5 mL tube. Incubate the samples overnight on a rocking device (*see* **Notes 16** and **17**).

3.4 3D Imaging

1. Place a SeeDB-processed sample on a 35 × 60 mm cover glass.
2. Place another cover glass over the sample to hold the liver sample between the two cover glasses. It is important to place the cover glass and the sample in parallel. Avoid making air bubbles between the liver sample and the cover glass (Fig. 2).
3. Attach the slide to the observation stage of the microscope. Use the sample holding arm to attach the slide to the stage. Use a sticky tape to attach the slide to the stage if a sample holding arm is unavailable.

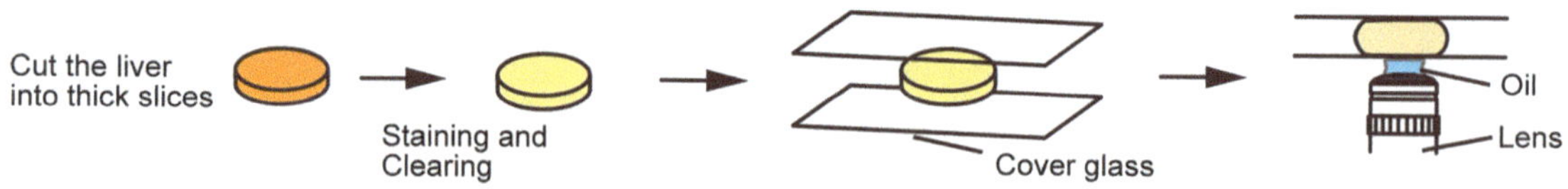

Fig. 2 Tissue processing workflow

4. It is highly recommended to use silicone or oil immersion lens to acquire images with high resolution (*see* **Note 18**). Place the oil between the lens and the sample slide. Make sure that the sample does not move when the stage moves.
5. Operate the confocal microscope according to the manufacturer's instructions. First, observe the fluorescent signal of labeled cells. The signal from a colony derived from a single hepatic progenitor is sparsely located (*see* **Note 19**). Set the acquisition range of the XY axis and then the acquisition range of the Z-axis. Set the step size of the Z-axis to the optimal minimum size of the objective lens. Set parameters for acquiring images (e.g., acquisition channels, sensitivity of the detectors, scanning speed, etc.). Start acquiring a 3D image of tdTomato and nuclei.
6. Repeat **step 5** to acquire the images of all colonies in a sample.
7. Change the sample, and acquire 3D images until enough data (>500 colonies) are collected.

3.5 Quantification

The 3D image is quantified using image processing software (Fig. 3). Here, we use Volocity (PerkinElmer) or ImageJ (Fiji) to quantify the images (*see* **Note 20**).

3.5.1 Quantification by Volocity

1. Load images using the software. All metadata will be imported automatically along with image data.
2. In Measurement mode, detect nuclei using the Find Object module. The object identified by this step includes all nuclei in the liver sample (Fig. 3a).
3. Detect labeled cells by applying the Find Object module to the tdTomato signal. tdTomato is uniformly localized throughout the cytoplasm, showing whole cells of the labeled colony (Fig. 3b).
4. Add the Filter population module, and set minimum object size to remove small objects resulting from noise.
5. Add the Separate touching object module into nuclei objects to separate incorrectly connected nuclei. This process is necessary because, in the first step of object identification, two different nuclei might be mistakenly identified as a single object.
6. Use the Intersect module with nuclei and cell objects. Nuclei of the labeled colony will be detected (Fig. 3c) (*see* **Note 21**).

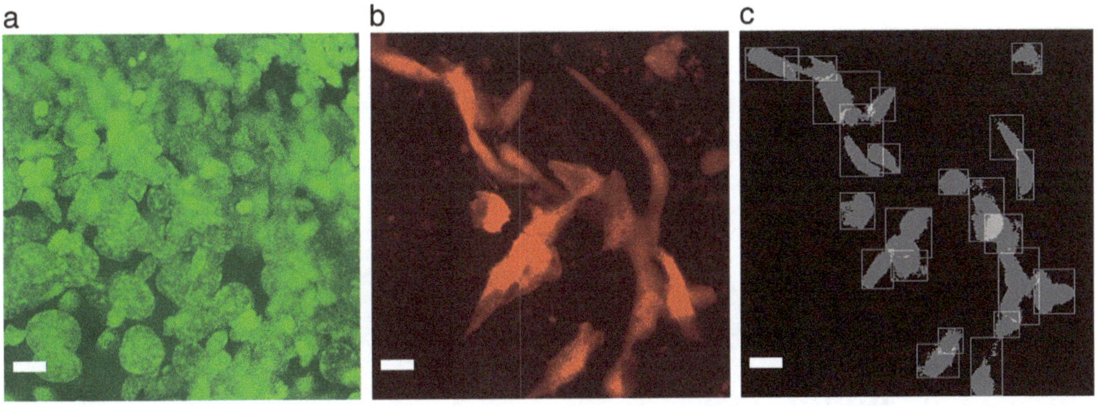

Fig. 3 Quantification process of three-dimensional (3D) images. A 3D image is projected on the 2D plane. (**a**) Nuclei. (**b**) Labeled colony visualized with tdTomato. (**c**) Identified nuclei of tdTomato-positive cells

7. After following this measurement protocol, you can apply the analytic protocol to many images all at once to quantify a large body of data.
8. Quantified data (Fig. 4) can be used in the statistical analysis (*see* **Note 22**).

3.5.2 Quantification by ImageJ (Fiji)

Instead of Volocity, we can use ImageJ/Fiji to quantify the images.

An ImageJ plug-in, MorphoLibJ, is required here. Please install Fiji (https://fiji.sc) [17] and MorphoLibJ (https://github.com/ijpb/MorphoLibJ/releases) [18] following manufacturer's instructions.

1. Load 3D image of nuclei and tdTomato to ImageJ. Serial tiff images can be imported directly ("File" → "Import" → "Image Sequence..."). Alternatively, 3D image files in another format can be loaded via plug-ins. For example, Olympus image format can be loaded by "Olympus ImageJ Plugin (https://imagej.net/OlympusImageJPlugin)."
2. Apply the Gaussian filter to both nuclei channel and tdTomato channels ("Process" → "Filters" → "Gaussian Blur").
3. Create binary images by setting the fluorescent intensity threshold ("Image" → "Adjust" → "Threshold" (method, default; dark background, on)). Please adjust threshold value to properly detect the objects. The optimal threshold value may change depending on various factors.
4. Merge nuclei and tdTomato channel ("Process" → "Image Calculator" (operation, AND)). Set the binary images of nuclei and tdTomato as input images.
5. Separate touching nuclei by the watershed method ("Plugins" → "MorphoLibJ" → "Binary Images" → "Distance Transform Watershed 3D" (distances, quasi-Euclidean;

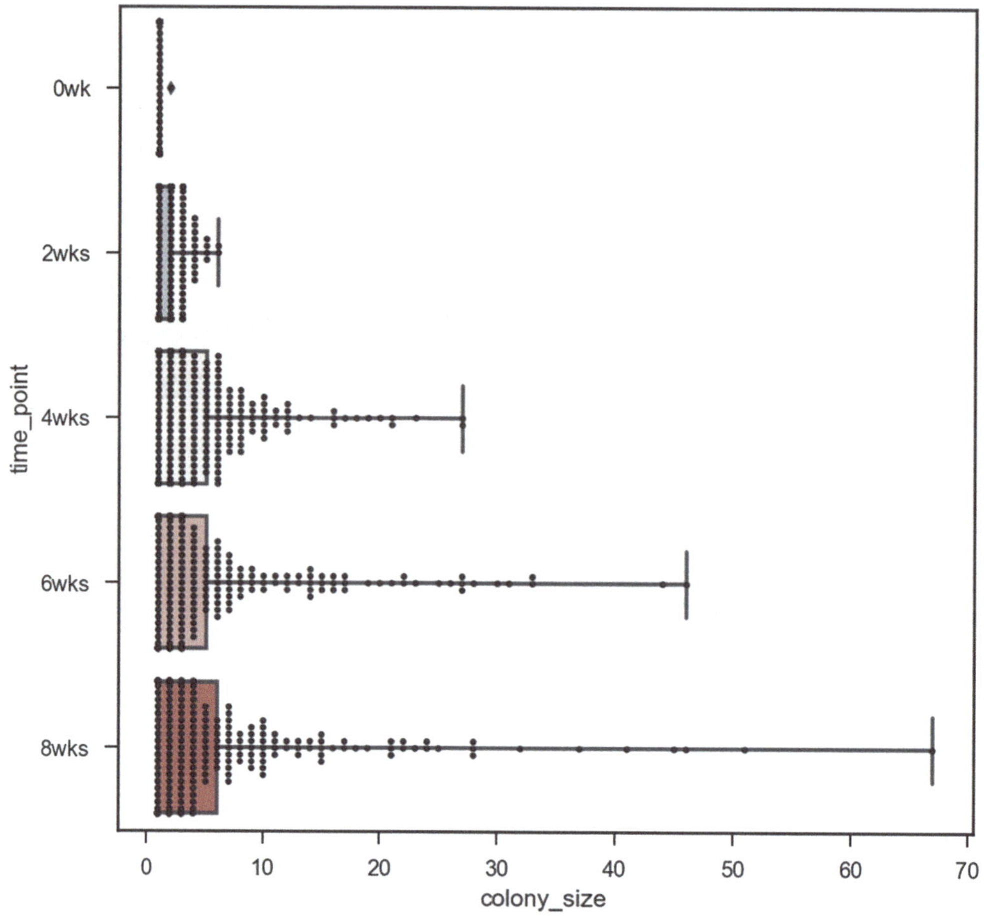

Fig. 4 Colony size distribution. Liver was analyzed after 0, 2, 4, 6, and 8 weeks of thioacetamide (TAA) injury. Data are shown as swarm plot and boxplot. Colony size distribution data were reproduced from our previous study [9]

normalize weights, off; dynamic, 3.00; connectivity, 26)). Segmented 3D images will be produced by this process.

6. In the segmented 3D image, each nucleus has unique label, but it has not been visualized yet. Visualize these labels by assigning a unique color to each label ("Plugins" → "MorphoLibJ" → "Label Images" → "Labels to RGB").
7. Visualize result in 3D view ("Plugins" → "3D viewer"). Make sure nuclei of tdTomato-positive cells were appropriately identified. If nuclei identification and segmentation failed, go back to the segmentation process (**step 5**). The extent of watershed segmentation can be optimized by the parameter of "Dynamic."
8. Output quantified values. First, select the segmented 3D image (produced by **step 5**), and then use the analysis function of MorphoLibJ

("Plugins" → "MorphoLibJ" → "Analyze" → "Particle Analysis 3D"). This procedure outputs quantified data table. The data still contain small non-nucleus objects made by noise. These noise objects can be removed by thresholding object volume.

4 Notes

1. The method utilizes an inducible Cre-loxP system to label and trace hepatic progenitor cells under in vivo conditions. Although we used the Prom1-CreERT2;R26R-tdTomato mouse strain, many additional mouse strains are available for lineage tracing of hepatic progenitor cells [10]. A different combination of Cre driver and fluorescent reporter strains can be used rather than the strains described here. Labeling efficiency might change depending on both the Cre driver and reporter strains. In addition, optimization of tamoxifen dosage is needed if a mouse strain different from Prom1-CreERT2; R26R-tdTomato is used.
2. TAA solution slowly degrades even when stored at 4 °C. Alternatively, a TAA stock solution (300 mg/30 mL) may be prepared and stored at −20 °C in 30 mL aliquots. Dilute one aliquot of the TAA stock solution (30 mL) in 1 L of water immediately before use.
3. SYTOX Green nucleic acid stain is used to visualize nuclei of liver tissue. The reagent emits a strong green fluorescent signal when bond to DNA. It can be used at very low concentrations, reducing experimental costs. Hoechst33342, another fluorescent reagent, can also be used for nucleus staining, but the staining protocol should be optimized for each reagent.
4. SeeDB is a tissue-clearing reagent containing D-fructose at a very high concentration [14]. D-fructose in SeeDB will easily precipitate if stored at 4 °C.
5. Image resolution and depth are highly dependent on the microscope and lens. The Olympus confocal microscope is used in this protocol, but another microscope may be used instead. Multi-photon microscopes can also be used.
6. The mouse age should be selected according to the experimental design. When examining hepatic progenitor cells at the developmental stage, 3–4-week-old mice can be used.
7. Tamoxifen dosage is critical for sparse cell labeling. At this dosage, approximately 0.2% of Prom1-positive hepatic progenitor cells will be labeled [9]. The amount of tamoxifen should be optimized if another mouse strain is used.

8. Intraperitoneal injection can be used instead of oral gavage with the same amount of tamoxifen.
9. In this protocol, a chronic liver injury induced by TAA was used to induce hepatic progenitor cell proliferation. There are also many other liver injury models that activate hepatic progenitors [3]. The 3,5-diethoxycarbonyl-1,4-dihydrocollidine diet, carbon tetrachloride, choline-deficient, ethionine-supplemented dietary, and the genetic liver injury models may be used [19]. However, hepatic progenitor cells exhibit various phenotypes depending on the type of injury, and the duration of liver injury varies depending on the model [4]. Therefore, the protocol should be optimized for each model.
10. Robust tissue fixation is required to keep tissue structure for later processing protocols. The entire tissue processing step takes at least 2 days. Liver tissue will be damaged by the process if fixation is incomplete. Fixation with 4% PFA is therefore recommended. Hydrophobic reagents such as paraffin, methanol, or acetone should not be used for fixation because they reduce the fluorescent signal of tdTomato.
11. Perfusion is strongly recommended for fixation for two reasons. First, perfusion is needed to drain blood from the liver tissue. The liver contains a large amount of blood, which scatters and absorbs the fluorescence signal. Any remaining blood deteriorates image quality. Second, perfusion is needed to inject PFA into the deeper areas of liver tissue. The liver has a very thick 3D structure, and perfusion is required to efficiently spread PFA throughout the organ. The color and hardness of the liver should be monitored when performing a perfusion.
12. It is important to control the temperature for a smooth cut of the samples. If temperature is too low, the liver becomes fragile and cracks easily.
13. Many liver sections are prepared by repeating this step to acquire images of labeled colonies enough for statistical analysis. Because of the sparse labeling, the small number of labeled hepatic progenitor cell colonies present in each section. Prepare many liver slices at a time. Sliced liver samples can be stored in the sample storage solution for up to 3 weeks. The experiment may be interrupted temporarily at this step.
14. If there are more than 100 samples, scale up the experimental design, and use 15 mL or 50 mL tubes instead of a 5 mL tube.
15. The liver section may break into small pieces if fixation is incomplete. Insufficiency of PFA fixation may result from incomplete perfusion of 4% PFA.
16. Do not incubate SeeDB at a temperature lower than 25 °C. Samples in SeeDB should be incubated at 37 °C before

microscopic examination. D-fructose in SeeDB precipitates easily, which deteriorates image quality. The sample is stable for several days, but the fluorescent protein signal slowly decays in SeeDB.

17. Make sure that the samples became transparent after processing with SeeDB. Another hydrophilic tissue-clearing reagent such as ScaleA2 [20] can be used to render the liver sample clear. In that case, however, it is necessary to optimize protocol details such as incubation time. A hydrophobic tissue-clearing reagent should not be used as it easily denatures fluorescent proteins. The clearing process is very critical for 3D imaging in the proposed method.
18. Use silicone or oil immersion lens as silicone oil reduces undesired refraction of excitation light. Random refraction should be minimized for deep imaging. The best magnification for the objective lens is around 30.
19. If a slide has too many colonies, the sparse labeling failed. The density of labeled hepatic progenitor cell colonies should be less than ten per liver slice. If not, the tamoxifen concentration should be lowered.
20. Many image processing software/plug-ins are available. Another software can be used instead of the protocol described here.
21. Hepatic progenitor cells have a high nucleus/cytoplasmic ratio and are located very close to each other in the colony. Image processing software sometimes fails to distinguish the individual nuclei. Therefore, it is recommended to verify the quantification results manually. Failure to obtain image quantification even after parameter optimization may result from insufficient imaging resolution or an insufficiently low signal/noise ratio.
22. You can infer much hidden information by analyzing quantified data using statistical methods [9, 11–13]. Although the count data itself may not be easily interpretable, there must be a biological mechanism behind the data. By analyzing the distribution of occurrence probability of a phenomenon, it is possible to estimate what kind of mechanism exists. The mechanism behind the distribution of count data is estimated by fitting the count data to various kinds of the probability distribution for the best fit.

    In the case where the known probability distribution cannot explain it, a new theoretical model is needed. The quantitative data acquisition method shown here brings further useful information in combination with such mathematical modeling and probability distribution.

References

1. Donati G, Watt FM (2015) Stem cell heterogeneity and plasticity in epithelia. Cell Stem Cell 16:465–476. https://doi.org/10.1016/j.stem.2015.04.014
2. Etzrodt M, Endele M, Schroeder T (2014) Quantitative single-cell approaches to stem cell research. Cell Stem Cell 15:546–558. https://doi.org/10.1016/j.stem.2014.10.015
3. Miyajima A, Tanaka M, Itoh T (2014) Stem/progenitor cells in liver development, homeostasis, regeneration, and reprogramming. Cell Stem Cell 14:561–574. https://doi.org/10.1016/j.stem.2014.04.010
4. Jelnes P, Santoni-Rugiu E, Rasmussen M et al (2007) Remarkable heterogeneity displayed by oval cells in rat and mouse models of stem cell-mediated liver regeneration. Hepatology 45:1462–1470. https://doi.org/10.1002/hep.21569
5. Suzuki A, Sekiya S, Onishi M et al (2008) Flow cytometric isolation and clonal identification of self-renewing bipotent hepatic progenitor cells in adult mouse liver. Hepatology 48:1964–1978. https://doi.org/10.1002/hep.22558
6. Okabe M, Tsukahara Y, Tanaka M et al (2009) Potential hepatic stem cells reside in EpCAM+ cells of normal and injured mouse liver. Development 136:1951–1960. https://doi.org/10.1242/dev.031369
7. Dorrell C, Erker L, Schug J et al (2011) Prospective isolation of a bipotential clonogenic liver progenitor cell in adult mice. Genes Dev 25:1193–1203. https://doi.org/10.1101/gad.2029411
8. Ema H, Morita Y, Yamazaki S et al (2006) Adult mouse hematopoietic stem cells: purification and single-cell assays. Nat Protoc 1:2979–2987. https://doi.org/10.1038/nprot.2006.447
9. Kamimoto K, Kaneko K, Kok C et al (2016) Heterogeneity and stochastic growth regulation of biliary epithelial cells dictate dynamic epithelial tissue remodeling. elife 5:e15034. https://doi.org/10.7554/eLife.15034.001
10. Lemaigre FP (2015) Determining the fate of hepatic cells by lineage tracing: Facts and pitfalls. Hepatology 61:2100–2103. https://doi.org/10.1002/hep.27659
11. Clayton E, Doupé DP, Klein AM et al (2007) A single type of progenitor cell maintains normal epidermis. Nature 446:185–189. https://doi.org/10.1038/nature05574
12. Doupé DP, Klein AM, Simons BD, Jones PH (2010) The ordered architecture of murine ear epidermis is maintained by progenitor cells with random fate. Dev Cell 18:317–323. https://doi.org/10.1016/j.devcel.2009.12.016
13. Driessens G, Beck B, Caauwe A et al (2012) Defining the mode of tumour growth by clonal analysis. Nature 488:527–530. https://doi.org/10.1038/nature11344
14. Ke M-T, Fujimoto S, Imai T (2013) SeeDB: a simple and morphology-preserving optical clearing agent for neuronal circuit reconstruction. Nat Neurosci 16:1154–1161. https://doi.org/10.1038/nn.3447
15. Zhu L, Gibson P, Currle DS et al (2009) Prominin 1 marks intestinal stem cells that are susceptible to neoplastic transformation. Nature 457:603–607. https://doi.org/10.1038/nature07589
16. Madisen L, Zwingman TA, Sunkin SM et al (2010) A robust and high-throughput Cre reporting and characterization system for the whole mouse brain. Nat Neurosci 13:133–140. https://doi.org/10.1038/nn.2467
17. Schindelin J, Arganda-Carreras I, Frise E et al (2012) Fiji: an open-source platform for biological-image analysis. Nat Publ Group 9:676–682. https://doi.org/10.1038/nmeth.2019
18. Legland D, Arganda-Carreras I, Andrey P (2016) MorphoLibJ: integrated library and plugins for mathematical morphology with ImageJ. Bioinformatics 32:3532–3534. https://doi.org/10.1093/bioinformatics/btw413
19. Forbes SJ, Newsome PN (2016) Liver regeneration - mechanisms and models to clinical application. Nat Rev Gastroenterol Hepatol 13:473–485. https://doi.org/10.1038/nrgastro.2016.97
20. Hama H, Kurokawa H, Kawano H et al (2011) Scale: a chemical approach for fluorescence imaging and reconstruction of transparent mouse brain. Nat Neurosci 14:1481–1488. https://doi.org/10.1038/nn.2928

Chapter 8

Chemical Screening Using a Zebrafish Model for Liver Progenitor Cell-Driven Liver Regeneration

Sungjin Ko and Donghun Shin

Abstract

Following massive hepatocyte ablation in zebrafish, biliary epithelial cells can extensively give rise to hepatocytes through liver progenitor cells (LPCs). The zebrafish liver injury model is an important system to elucidate the molecular mechanisms underlying LPC-driven liver regeneration. Here, we describe a chemical screening method using the zebrafish model for identifying small molecules that can modulate LPC-driven liver regeneration.

Key words Nitroreductase, Biliary epithelial cells, Hepatocyte ablation, Chemical screen, Liver progenitor cells

1 Introduction

The liver is a highly regenerative organ: it can recover even after losing 70% of its mass [1]. However, after prolonged or severe liver injury, as observed in the advanced stages of chronic and acute liver diseases, the liver loses its ability to regenerate through self-replication of hepatocytes [2, 3]. This continuing damage in combination with defective hepatocyte proliferation stimulates excessive inflammation and subsequent wound healing mechanisms, including fibrotic/cirrhotic responses [1–3]. Currently, liver transplantation is the only reliable cure to prolong the life of patients with advanced liver diseases; however, the number of liver donors is not sufficient to match the increasing number of patients [4]. To alleviate this bottleneck and to extend patient lives until transplantation, diverse approaches have been attempted to develop therapeutic options [5–7]. Although there have been technical improvements, further progress is still needed for these therapeutics to be considered as reliable options worthy of testing in clinical trials.

One attractive approach that has recently garnered attention is to promote the differentiation of activated liver progenitor cells (LPCs) in diseased livers into either hepatocytes or biliary epithelial

Naoki Tanimizu (ed.), *Hepatic Stem Cells: Methods and Protocols*, Methods in Molecular Biology, vol. 1905, https://doi.org/10.1007/978-1-4939-8961-4_8, © Springer Science+Business Media, LLC, part of Springer Nature 2019

cells (BECs) [3, 8]. Given the prevalence of LPCs in diseased human livers [2, 9], this approach becomes more promising due to recent findings in mice that activated BEC-derived LPCs are a key source of new hepatocytes in various liver injury settings with defective hepatocyte proliferation [9, 10]. While these achievements are encouraging to elucidate the molecular mechanisms underlying LPC-driven liver regeneration and to identify drug targets that can augment this process, the mouse is not an ideal model animal for large-scale chemical screening due to its high cost and time constraint.

Given the ease of chemical screening in zebrafish embryos and larvae [11], we have established a zebrafish liver injury model for LPC-driven liver regeneration by generating a transgenic line, *Tg (fabp10a:CFP-NTR)*, that provides a genetic means to ablate differentiated hepatocytes [12]. The transgenic fish expresses bacterial nitroreductase (NTR) enzymes fused with cyan fluorescent proteins (CFP) specifically in hepatocytes due to the hepatocyte-specific *fabp10a* promoter. Since NTR converts the non-toxic prodrug metronidazole (Mtz) into a cytotoxic drug, treating the fish with Mtz ablates only NTR-expressing hepatocytes. Following massive hepatocyte ablation, BECs dedifferentiate into LPCs and subsequently differentiate into hepatocytes [12]. Compared to the aforementioned mouse model, this zebrafish model is advantageous for chemical screening to identify small molecules that affect LPC-driven liver regeneration [13]: (1) the entire liver regeneration assay takes only 6 days, (2) the regeneration process is homogeneous with very small variability among animals, and (3) hundreds of animals can be manipulated simultaneously. The zebrafish assay allows for screening ~30 compounds per week. Here we describe a detailed protocol for chemical screening using the zebrafish hepatocyte ablation model.

2 Materials

1. *Tg(fabp10a:CFP-NTR);Tg(fabp10a:DsRed)* adult fish.
2. Mating tanks: large aquaculture tanks with dividers.
3. Fine plastic mesh strainer (tea strainer).
4. Squeeze bottle.
5. Petri dishes: 100 × 15 mm and 60 × 15 mm.
6. Egg water: 0.3 g of sea salts "Instant Ocean" in 1 L of distilled water supplemented with 0.5-mM $CaSO_4$.
7. Disposable polyethylene transfer pipettes.
8. Stereomicroscope.

9. 5-mM phenylthiourea (PTU) stock solution: 760 mg of PTU in 1 L of distilled water. Heat the solution to 60 °C until completely dissolved.
10. 24.2-mM tricaine (3-amino benzoic acid ethyl ester, also known as MS-222) stock solution: 400 mg of tricaine powder in 100 mL of distilled water, pH 7–7.5.
11. 100% dimethyl sulfoxide (DMSO, molecular grade).
12. Epifluorescence microscope with DsRed and CFP filters.
13. 10-mM Mtz solution: 0.171 g of Mtz in 100 mL of egg water supplemented with 0.2% DMSO and 0.2-mM PTU. Stir rapidly using a magnetic stirrer at room temperature (RT) for 30 min.
14. 12-well tissue culture plates.
15. 15-mL conical tubes.
16. Aluminum foil.
17. 3% methyl cellulose: add 15 g of methyl cellulose into 500 mL of heated egg water (60 °C) and stir with a magnetic stirrer at RT for 20 min.
18. Superfine eyelash with handle (Ted Pella Inc., Prod No. 113).
19. ImageJ software: it can be obtained at https://imagej.nih.gov/ij/index.html.
20. Nitrile gloves.
21. System water: water used to raise and maintain zebrafish.

3 Methods

The outline of the screening process is shown in Fig. 1.

3.1 Preparation of Larvae

1. In each mating tank, place *Tg(fabp10a:CFP-NTR);Tg (fabp10a:DsRed)* male (3) and female (3) fish that are separated by sex with a tank divider. To obtain enough embryos for screening, set up five tanks. Keep tanks at 28 °C overnight.
2. On the following morning, replenish the tanks with fresh system water, and remove the dividers. After 1–2 h, collect embryos by straining the water containing eggs with a fine plastic mesh strainer.
3. Rinse the strainer with a fine stream of system water from a squeeze bottle. Transfer embryos into 100-mm petri dishes with ~25 mL of fresh egg water. Keep no more than 100 embryos per dish and incubate them at 28 °C.
4. About 5–6 h later, under a stereomicroscope, remove unfertilized embryos using a transfer pipette. Replenish with fresh egg water and keep at 28 °C overnight.

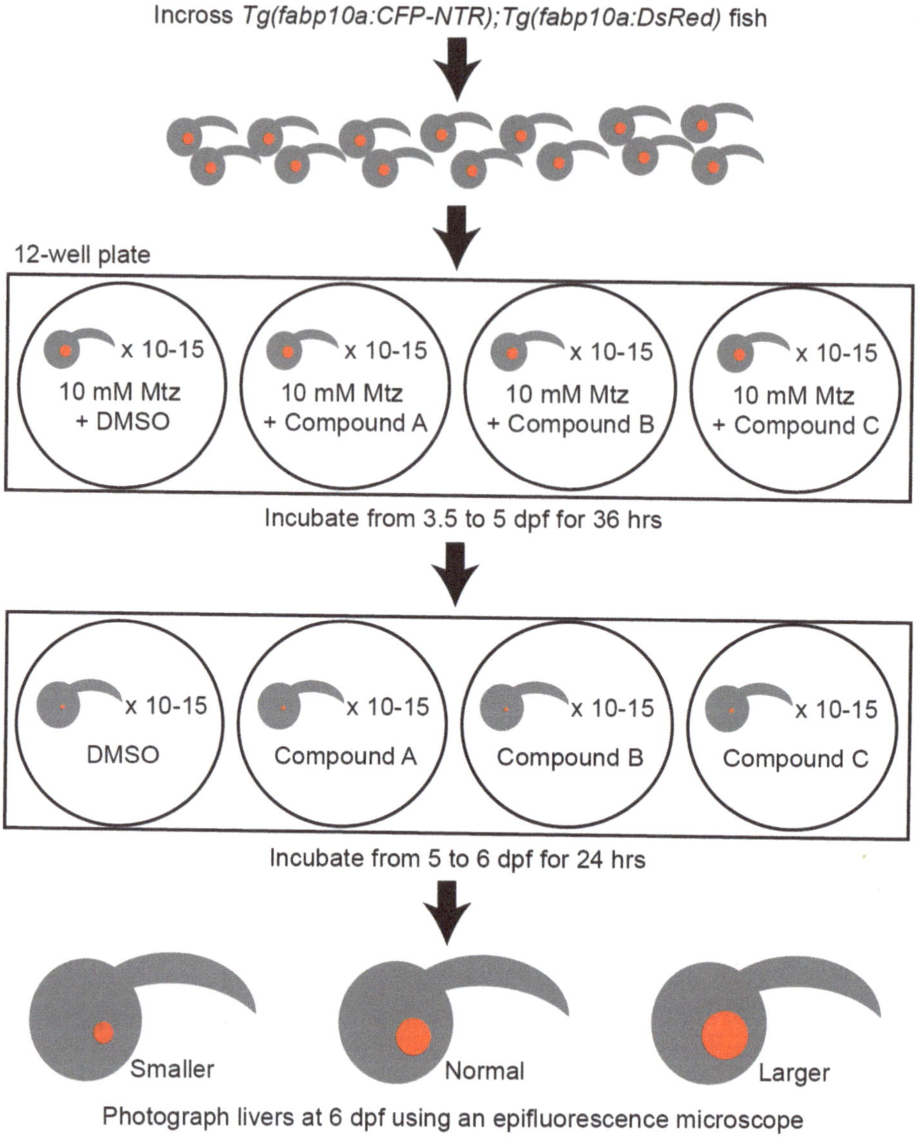

Fig. 1 Cartoon illustrating the entire chemical screening process for identifying compounds that affect LPC-driven liver regeneration. Red circles denote livers

5. On the next morning around 24-h post-fertilization (hpf), check the embryos under a stereomicroscope, and remove dead or deformed embryos using a transfer pipette. After cleaning, to inhibit embryo pigmentation, add PTU (final concentration: 0.2 mM) into each petri dish, and raise embryos until 3.3 days post-fertilization (dpf) at 28 °C (*see* **Note 1**).
6. Anesthetize the embryos by adding tricaine (final concentration: 0.016%) into the petri dishes (*see* **Note 2**). Sort CFP and DsRed double-positive larvae under an epifluorescence

microscope. Transfer the double-positive embryos into a new 100-mm petri dish. Remove the egg water containing tricaine and wash three times with egg water. Then, add fresh egg water supplemented with 0.2-mM PTU and keep the embryos at 28 °C.

3.2 Mtz and Compound Treatments

1. Prepare 10-mM Mtz solution (*see* Subheading 2, **item 13,** and **Note 3**).
2. Prepare the mixture of each compound and Mtz in a 15-mL conical tube: add 6 μL of each compound solution (*see* **Note 4**) into 2-mL Mtz solution and vortex (*see* **Note 5**). For a control group, add 6 μL of DMSO into 2-mL Mtz solution (*see* **Note 6**).
3. Dispense 10–15 larvae into each well of 12-well tissue culture plates using a transfer pipette (*see* **Note 5**). Remove egg water from each well and add 2 mL of Mtz/compound mixture. To prevent the photoinactivation of Mtz and compounds, cover the plates with aluminum foil, and incubate them at 28 °C for 36 h.
4. Prepare each compound solution in egg water: add 6 μL of each compound solution and 4 μL of DMSO into 2 mL of egg water supplemented with 0.2-mM PTU and vortex. For a control group, add 10 μL of DMSO into 2 mL of egg water supplemented with 0.2-mM PTU (*see* **Note 5**).
5. After 36 h of incubation, to immobilize the larvae, add a lower concentration of tricaine (final concentration: 0.004%) (*see* **Note** 7). Sort out larvae with well-ablated livers (optional, *see* **Note 8**). Remove the Mtz/compound solution and add 2 mL of egg water. Gently transfer the embryos into a well of a new 12-well plate using a transfer pipette and swirl the plate twice to wash the embryos.
6. Remove the egg water and add 2 mL of the compound-containing egg water. Cover the plates with aluminum foil and incubate at 28 °C for 24 h.

3.3 Image Acquisition and Analysis

1. Anesthetize the larvae by adding tricaine (final concentration: 0.016%) into each well of the 12-well plates. Gently transfer the larvae into 60-mm petri dishes filled with 3% methyl cellulose and using a superfine eyelash (*see* Subheading 2, **item 18**); orient the larvae laterally to clearly reveal the liver.
2. Take a picture using an epifluorescence microscope with a DsRed filter (Fig. 2).
3. Once images are acquired, the ImageJ software is used to measure liver area, the integrated density of DsRed, and the mean background fluorescence, which are used to calculate corrected total cell fluorescence (CTCF) using the formula

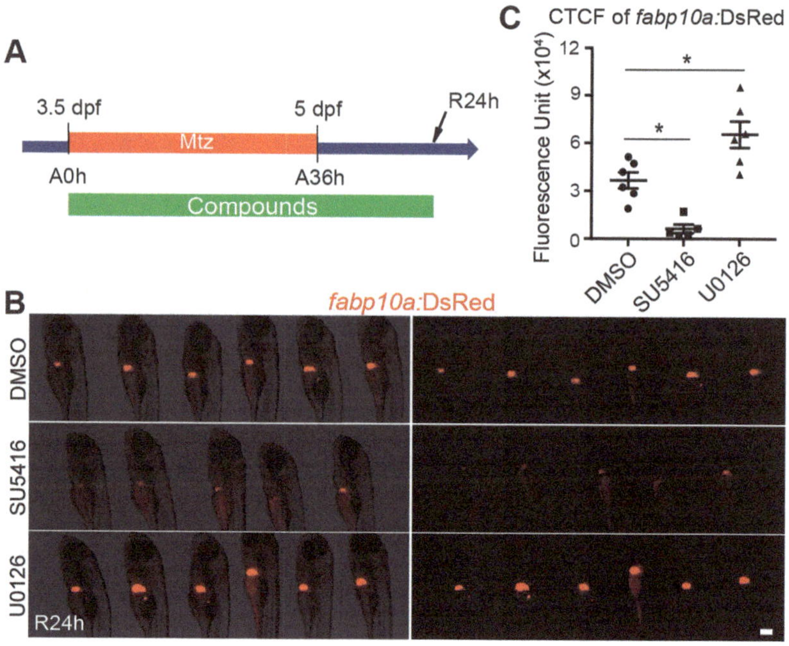

Fig. 2 Chemical screening identifies SU5416 and U0126 as a negative and a positive regulator of LPC-driven liver regeneration, respectively. (**a**) Experimental scheme illustrating the stages of Mtz and compound treatments and analysis (arrow). A0h and A36h, ablation for 0 and 36 h, respectively; R24h, regeneration for 24 h. (**b**) Epifluorescence images showing *fabp10a*:DsRed expression in regenerating larvae at R24h. 2-μM SU5416 and 100-μM U0126 were used. Scale bar, 100 μm. (**c**) Quantification of *fabp10a*:DsRed expression in the regenerating livers, as shown in (**b**). One-way ANOVA; *, $P < 0.05$; error bars, ± SEM

below [14]. CTCF value represents DsRed expression level (Fig. 2c).

CTCF = Integrated density—(selected area × the mean fluorescence of background readings).

4 Notes

1. PTU is toxic. Use in accordance with appropriate handling guidelines.
2. Tricaine is toxic. Use in accordance with appropriate handling guidelines.
3. Prolonged exposure or increased concentration of Mtz is toxic. Use in accordance with appropriate handling guidelines.
4. The working concentration of each compound can be determined by treating *Tg(fabp10a:CFP-NTR)* larvae in a 96-well plate (three larvae per well) with various concentrations of the

compound together with 10-mM Mtz; the maximum tolerated concentration is used for screening.

5. For a well of a 12-well plate, 2 mL is a proper volume (10–15 larvae per well); 5 mL is for a well of a 6-well plate (15–25 larvae per well).
6. Most chemicals are dissolved in DMSO as a concentrated stock. To prevent any chemical hazards, nitrile gloves are recommended during all screening procedures.
7. Tricaine is metabolized in the liver and can be hepatotoxic; to avoid aggravating liver damage in the Mtz-treated larvae, use 1/4–1/5 dose of tricaine.
8. (Optional) Remove larvae with partial ablation by examining them under an epifluorescence microscope with a DsRed filter. Use larvae with a tiny liver, indicating the full ablation of hepatocytes. Caution: be aware that some compounds can suppress NTR/Mtz-induced hepatocyte ablation, preventing the full ablation. Thus, the effect of these compounds on LPC-driven liver regeneration cannot be assessed using the zebrafish hepatocyte ablation model.

Acknowledgments

We thank Jackie Russell, Mehwish Khaliq, and Michael Tsang for editing the manuscript. This study was supported by an NIH grant (DK101426) to D.S.

References

1. Michalopoulos GK (2013) Principles of liver regeneration and growth homeostasis. Compr Physiol 3(1):485–513. https://doi.org/10.1002/cphy.c120014
2. Boulter L, Lu WY, Forbes SJ (2013) Differentiation of progenitors in the liver: a matter of local choice. J Clin Invest 123(5):1867–1873. https://doi.org/10.1172/JCI66026
3. Forbes SJ, Newsome PN (2016) Liver regeneration - mechanisms and models to clinical application. Nat Rev Gastroenterol Hepatol 13(8):473–485. https://doi.org/10.1038/nrgastro.2016.97
4. Kim WR, Lake JR, Smith JM, Schladt DP, Skeans MA, Harper AM, Wainright JL, Snyder JJ, Israni AK, Kasiske BL (2018) OPTN/SRTR 2016 annual data report: liver. Am J Transplant 18(Suppl 1):172–253. https://doi.org/10.1111/ajt.14559
5. Ibars EP, Cortes M, Tolosa L, Gómez-Lechón MJ, López S, Castell JV, Mir J (2016) Hepatocyte transplantation program: lessons learned and future strategies. World J Gastroenterol 22(2):874–886. https://doi.org/10.3748/wjg.v22.i2.874
6. Garg V, Garg H, Khan A, Trehanpati N, Kumar A, Sharma BC, Sakhuja P, Sarin SK (2012) Granulocyte colony-stimulating factor mobilizes CD34(+) cells and improves survival of patients with acute-on-chronic liver failure. Gastroenterology 142(3):505–512.e1. https://doi.org/10.1053/j.gastro.2011.11.027
7. Alwahsh SM, Rashidi H, Hay DC (2017) Liver cell therapy: is this the end of the beginning? Cell Mol Life Sci 75(8):1307–1324. https://doi.org/10.1007/s00018-017-2713-8
8. Lukacs-Kornek V, Lammert F (2017) The progenitor cell dilemma: cellular and functional heterogeneity in assistance or escalation of liver injury. J Hepatol 66(3):619–630. https://doi.org/10.1016/j.jhep.2016.10.033

9. Raven A, Lu WY, Man TY, Ferreira-Gonzalez S, O'Duibhir E, Dwyer BJ, Thomson JP, Meehan RR, Bogorad R, Koteliansky V, Kotelevtsev Y, Ffrench-Constant C, Boulter L, Forbes SJ (2017) Cholangiocytes act as facultative liver stem cells during impaired hepatocyte regeneration. Nature 547(7663):350–354. https://doi.org/10.1038/nature23015
10. Lu WY, Bird TG, Boulter L, Tsuchiya A, Cole AM, Hay T, Guest RV, Wojtacha D, Man TY, Mackinnon A, Ridgway RA, Kendall T, Williams MJ, Jamieson T, Raven A, Hay DC, Iredale JP, Clarke AR, Sansom OJ, Forbes SJ (2015) Hepatic progenitor cells of biliary origin with liver repopulation capacity. Nat Cell Biol 17(8):971–983. https://doi.org/10.1038/ncb3203
11. Wiley DS, Redfield SE, Zon LI (2017) Chemical screening in zebrafish for novel biological and therapeutic discovery. Methods Cell Biol 138:651–679. https://doi.org/10.1016/bs.mcb.2016.10.004
12. Choi TY, Ninov N, Stainier DY, Shin D (2014) Extensive conversion of hepatic biliary epithelial cells to hepatocytes after near total loss of hepatocytes in zebrafish. Gastroenterology 146(3):776–788. https://doi.org/10.1053/j.gastro.2013.10.019
13. Ko S, Choi TY, Russell JO, So J, Monga SPS, Shin D (2015) Bromodomain and extraterminal (BET) proteins regulate biliary-driven liver regeneration. J Hepatol 64(2):316–325. https://doi.org/10.1016/j.jhep.2015.10.017
14. McCloy RA, Rogers S, Caldon CE, Lorca T, Castro A, Burgess A (2014) Partial inhibition of Cdk1 in G 2 phase overrides the SAC and decouples mitotic events. Cell Cycle 13(9):1400–1412. https://doi.org/10.4161/cc.28401

Part III

Generation of Hepatocytes, Cholangiocytes, and Their Progenitors

Chapter 9

Conversion of Fibroblasts to Hepatocytes In Vitro

Pengyu Huang, Lulu Sun, Ludi Zhang, and Lijian Hui

Abstract

Primary hepatocytes are widely used in regenerative medicine, drug metabolism analysis, and in vitro drug screens. To overcome the shortage of liver donors, several strategies, such as differentiation of pluripotent stem cells and transdifferentiation from somatic cells, were developed to generate hepatocytes from alternative sources. Here, we describe in detail lenti-virus-based procedure for direct conversion of human fibroblasts to hepatocytes (hiHep cells) in vitro. A detailed protocol for preparation of human fibroblasts from scar tissues is also provided. Based on this protocol, *FOXA3*, *HNF1A*, and *HNF4A* are introduced into SV40-large-T-antigen-expressing human scar fibroblasts by lenti-virus. It usually takes about 5–7 days to get epithelial hiHep colonies. SV40-large-T-antigen-expressing hiHep (hiHep^{LT}) cells are proliferative and can be expanded to a large number for potential uses.

Key words Hepatocytes, HiHep cells, Transdifferentiation, Reprogramming

1 Introduction

Most mammalian somatic cells that retain integrated genomes are supposed to be plastic and could be converted to other cell lineage after reprogramming. Transdifferentiation, a process of conversion of one differentiated cell type to another, thus attracts a lot of interest for its application potential in generating desired cell types for regenerative medicine. For decades, many efforts have been made to generate hepatocytes or hepatocyte-like cells by transdifferentiation. Pancreatic cell, a lineage closely related to the liver during development, was firstly used as the starting cell type to generate hepatocytes in vivo and in vitro [1–4]. Recently, more accessible cell types, such as mesenchymal stem cell and fibroblast, showed great potential as the starting cell types for transdifferentiation. Previously, we have reported the successful conversion of mouse and human fibroblasts to hepatocytes by enforced expression of several hepatic transcription factors [5, 6]. These human-induced hepatocytes (hiHep cells) expressed many hepatic marker genes and acquired several hepatic functions, such as glycogen

Naoki Tanimizu (ed.), *Hepatic Stem Cells: Methods and Protocols*, Methods in Molecular Biology, vol. 1905, https://doi.org/10.1007/978-1-4939-8961-4_9, © Springer Science+Business Media, LLC, part of Springer Nature 2019

storage, secretion of albumin, and metabolism of drugs. The $hiHep^{LT}$ cells also showed the ability of integration into immune-deficient mouse livers and functions in vivo [6].

In this chapter, we provided an example of protocol to convert human fibroblasts to expandable $hiHep^{LT}$ cells. The human fibroblasts are generated from human scar tissues and immortalized by enforced expression of SV40 large T antigen before transdifferentiation. SV40-large-T-antigen-expressing fibroblasts are further infected with lenti-virus expressing FOXA3, HNF1A, and HNF4A. The medium is then exchanged to hiHep cell culture medium supplemented with 1% fetal bovine serum (FBS) 2 days after infection. Usually, epithelial colonies form on the dishes after 5–7 days' induction (Fig. 1). Co-expression of albumin and AAT is an important feature of hepatocytes and is usually used to evaluate the efficiency of generation of $hiHep^{LT}$ cells. Around 10–40% of SV40-large-T-antigen-expressing fibroblasts can be converted to albumin and AAT double-positive cells in most cases (Figs. 2 and 3). Compared to $hiHep^{LT}$ cells, fibroblasts are more sensitive to collagenase digestion. Thus we can use collagenase to eliminate fibroblasts and enrich $hiHep^{LT}$ cells (Figs. 2 and 3).

Human fibroblasts derived from fetal tissue, foreskin, and scar tissue are usually used as the starting cells due to their easy accessibility and proliferative potential. Primary human fibroblasts are capable of proliferation for 20–30 passages under recommended culture conditions. However, hiHep cells generated from human fibroblasts usually fail to proliferate. Thus it is not applicable to get a large number of hiHep cells. To overcome this issue, we introduced SV40 large T antigen into human fibroblasts before transdifferentiation. $HiHep^{LT}$ cells derived from SV40-large-T-antigen-expressing fibroblasts acquired the ability to proliferate. Expansion of $hiHep^{LT}$ cells to the number of around 3×10^9 has been achieved for hiHep-cell-based bioartificial liver device [7]. Importantly, $hiHep^{LT}$ cells of late passage still express hepatic genes and retain hepatic functions [7]. In addition to fibroblasts, other easy accessible cells, such as mesenchymal stem cells, exfoliated renal epithelial cells from urine, and umbilical cells, are also possible cell sources for hepatic transdifferentiation [8, 9].

Generation of hiHep cells without expression of SV40 large T antigen is also practicable with a similar protocol. However, hiHep cells are not proliferative. Only scattered epithelial cells and a few epithelial clusters could be found after 7–10 days' induction. Moreover, the transdifferentiation efficiency is largely affected by starting cell density and multiplicity of the infection (MOI) of the virus. Thus it is very important to perform pre-experiments to test the best starting cell density and MOI of the virus. Comparing to fibroblasts without expression of SV40 large T antigen, SV40-large-T-antigen-expressing fibroblasts are less sensitive to the starting cell density and MOI during induced hepatic

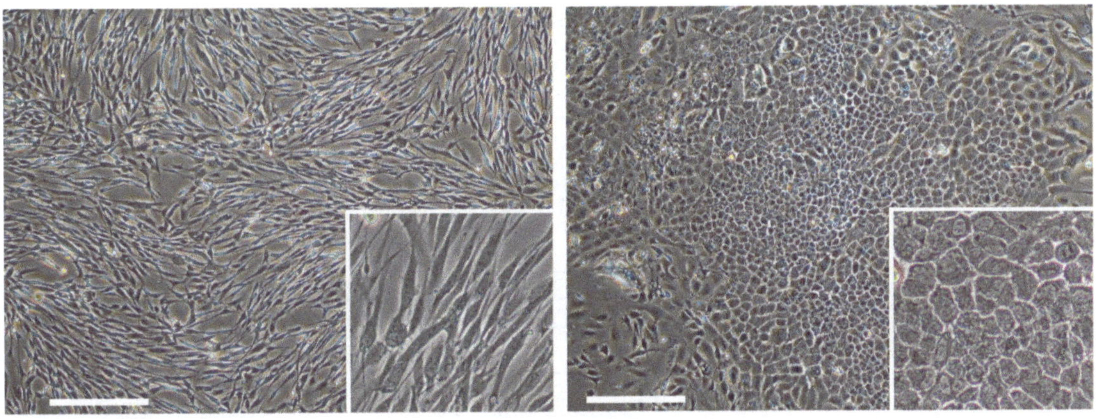

Fig. 1 Representative pictures of human scar fibroblasts and hiHep^{LT} cells. Primary human scar fibroblasts (passage 3) were generated from human scar tissue and infected with lenti-virus expressing SV40 large T antigen (left). HiHep^{LT} cells were converted from SV40-large-T-antigen-expressing human scar fibroblasts by enforced expressions of FOXA3, HNF1A, and HNF4A (right). Scale bar: 500 μm

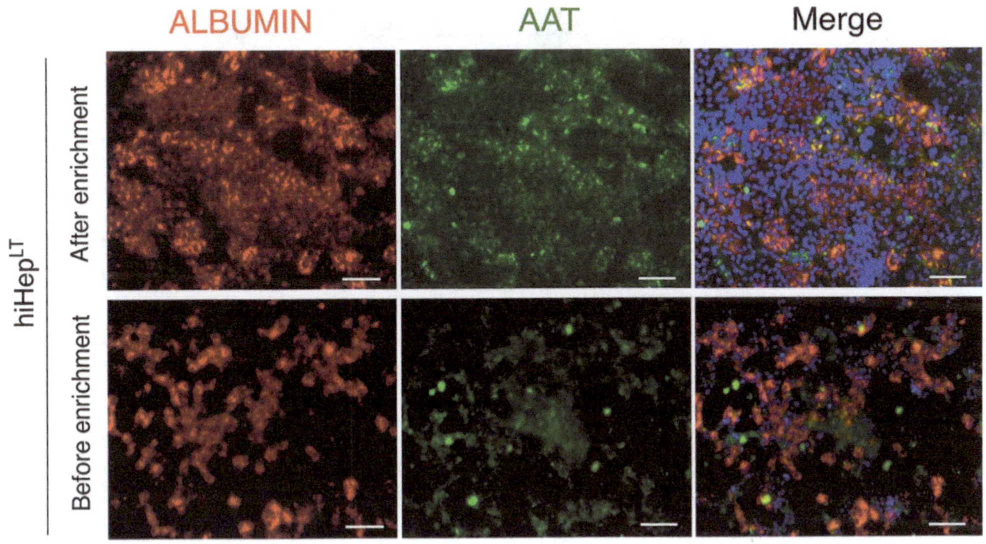

Fig. 2 Representative pictures of hiHep^{LT} cells stained with albumin and α-1-antitrypsin (AAT) before and after enrichment. Scale bar: 200 μm

transdifferentiation. It is highly recommended to start with SV40-Large-T-Antigen-expressing fibroblasts for researchers who have no experience in hepatic transdifferentiation.

The hepatic transcription factors used for hepatic transdifferentiation were optimized and minimized to FOXA3, HNF1A, and HNF4A in our previous study [6]. However, several studies also indicated that different combinations of transcription factors could achieve similar transdifferentiation processes [5, 6, 10]. After interrogation of the mechanisms of hepatic transdifferentiation,

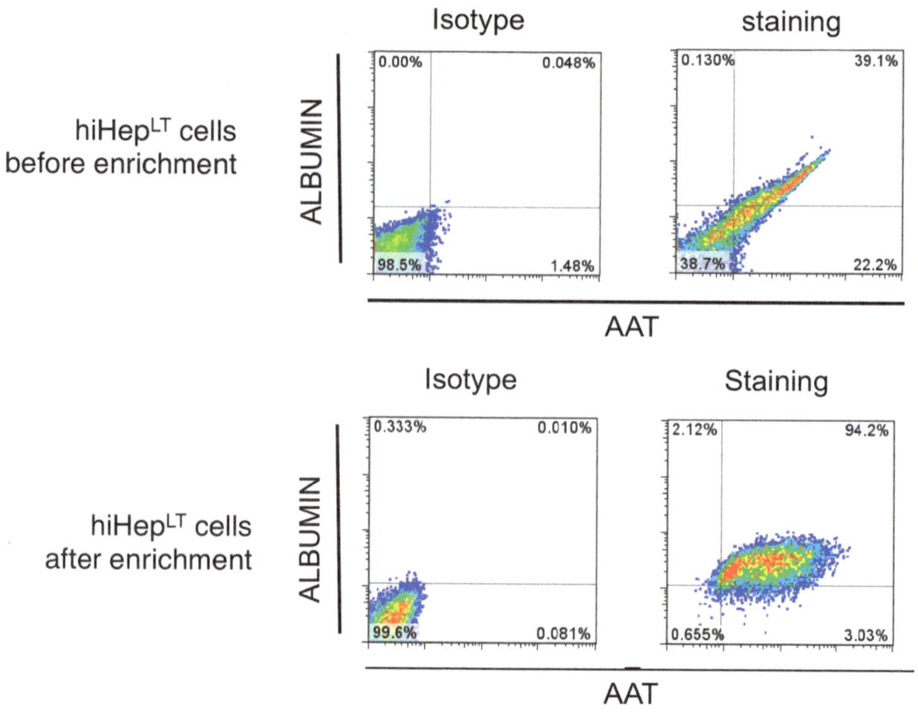

Fig. 3 Albumin and AAT double-positive hiHep^{LT} cells were quantified by flow cytometry before and after enrichment

inactivation or activation of certain signaling pathways may also be used for improvement of this process in the future [11].

2 Materials

2.1 Biological Material

Human scar tissues are taken from scar tissues removed during surgical operations after written consent following guidelines approved by an Ethic Committee.

2.2 Culture Media and Reagents

1. 293FT cells.
2. The human fibroblast culture (HFC) medium: DMEM/F12 1:1 mixture supplemented with 10% fetal bovine serum (FBS), recombinant human 5-ng/mL bFGF, 100-nM β-mercaptoethanol, 1× MEM non-essential amino acids (MEM NEAA), and 0.5-mM sodium pyruvate (*see* **Note 1**).
3. The hiHep cell culture medium: DMEM/F12 1:1 mixture supplemented with 40-ng/mL recombinant human TGF-α, 40-ng/mL recombinant human EGF, 10-μM dexamethasone, 1× insulin-transferrin-selenium (ITS), 2-g/L galactose, 0.1-g/L ornithine, 0.3-g/L proline, 0.6-g/L nicotinamide, 0.544-

mg/L $ZnCl_2$, 0.75-mg/L $ZnSO_4 \cdot 7H_2O$, 0.2-mg/L $CuSO_4 \cdot 5H_2O$, 0.025-mg/L $MnSO_4$.

4. 2-mg/mL collagenase IV solution: dissolve 20-mg collagenase IV in 10-mL PBS.
5. pWPI expression vector.
6. Collagen I solution: Dissolve rat tail-derived collagen I in 0.6% (v/v) acetic acid solution to 25 mg/L at 4 °C overnight.
7. Calcium phosphate solution: 7.5-μg psPAX2, 3-μg pMD2.G, 31-μL 2.5-M $CaCl_2$ in bi-distilled water. The total volume is adjusted to 250 μL.
8. 2× HEPES solution: 0.8-g NaCl, 0.027-g $Na_2HPO_4 \cdot 2H_2O$, and 1.2-g HEPES in 100-mL bi-distilled water. Calibrate the pH to 7.05.
9. 0.05% trypsin.
10. 8-mg/mL polybrene.

3 Methods

3.1 Coating Dish

1. Add collagen I solution to cell culture dishes (1.5-μg/cm^2 surface).
2. Air-dry the collagen I-coated dishes in a biosafety cabinet for future use.

3.2 Preparation of Lenti-Virus

1. CDNA of SV40 large T antigen gene, FOXA3, HNF1A, and HNF4A were inserted into a pWPI expression vector.
2. Seed 1 × 10^6 293FT cells on a 100-mm dish, and add 10-mL DMEM +10% FBS. Culture 293FT cells for 24 h.
3. One hour before transfection, exchange the culture medium with 6 mL of prewarmed DMEM+10% FBS.
4. Mix 250-μL calcium phosphate solution containing 10-μg pWPI expression vector and 250-μL 2× HEPES solution by pipetting. Keep the mixture for 5 min at room temperature then transfer the mixture to 293FT cells cultured in a 100-mm dish.
5. Incubate the cells for 6 h and exchange the medium with DMEM +10% FBS.
6. 48 h after transfection, collect supernatant and filtered by a 0.45-μm filter. Store the virus solution in −80 °C.

3.3 Preparation of Human Scar Fibroblasts

1. Wash human scar tissue with 70% ethanol three times, then with PBS once.
2. Cut the tissue into 2–5-mm^3 pieces.

3. Place three tissue pieces per 60-mm collagen I-coated dish, and incubate at 37 °C for 1 h in an incubator.
4. Carefully add 5 mL of HFC medium to the dish (*see* **Note 2**).
5. Put the dish in an incubator at 37 °C for 5 days without exchange of the medium, and then exchange the medium every 2 days.
6. Human scar fibroblasts migrate out of the tissue after 2–3 weeks (passage 1).
7. When the cells become confluent, remove HFC medium and the tissues.
8. Wash once with PBS and trypsinize with 1 mL of 0.05% trypsin for 15 min.
9. Suspend the cells with 2 mL of HFC medium. Centrifuge the cells at 150 × *g* for 5 min.
10. Discard the supernatant, resuspend the cells with 5 mL of HFC medium, and transfer to a 60-mm collagen I-coated dish (passage 2).
11. When the cells become confluent, passage the cells to new collagen I-coated dishes (5×10^5 cells per 60-mm collagen I-coated dish, passage 3). Exchange HFC medium every 2 days.
12. For the following passages, seed 5×10^5 cells per 60-mm collagen I-coated dish and exchange HFC medium every 2 days.

3.4 Immortalization of Human Scar Fibroblasts

1. Seed 5×10^5 human scar fibroblasts (passage 3–5) on a 60-mm collagen I-coated dish, and add HFC medium to 5 mL.
2. The next day, thaw lenti-virus expressing SV40 large T antigen on ice. Mix lenti-virus (MOI = 2), 5 μL of 8-mg/mL polybrene and HFC medium to the volume of 5 mL.
3. Remove the medium from the dish. Add 5 mL of lenti-virus, polybrene, and HFC medium mixture.
4. After 24 h infection, remove the medium and add 5 mL of fresh HFC medium.
5. When the cells become confluent, passage the cells to new collagen I-coated dishes (5×10^5 cells per 60-mm dish).
6. For the following passages, seed 5×10^5 cells per 60-mm collagen I-coated dish and exchange HFC medium every 2 days.

3.5 Conversion of HSF^{LT} to $hiHep^{LT}$ Cells

1. Expand human scar fibroblasts expressing SV40 large T antigen (HSF^{LT}) to 70–80% confluence.
2. Discard the medium and wash twice with 2 mL of PBS.

3. Trypsinize the cells with 1 mL of 0.05% trypsin at 37 °C for 5 min.
4. Add 2 mL of HFC medium to suspend the cells, and transfer the cells to a 15-mL tube. Count the number of cells.
5. Thaw lenti-virus expressing FOXA3, HNF1A, and HNF4A on ice.
6. Mix lenti-virus (MOI = 1.25 for each virus, *see* **Note 3**), 5 μL of 8-mg/mL polybrene, 2×10^5 HSF^{LT} (*see* **Note 3**), and HFC medium to the final volume of 5 mL. Mix gently by pipetting up and down. Transfer the mixture to a 60-mm collagen I-coated dish. Incubate the cells at 37 °C (day 0).
7. 24 h after infection, remove the medium and add fresh HFC medium (day 1).
8. The next day, remove the medium. Add 5 mL of hiHep cell culture medium with 1% FBS (day 2).
9. From day 4 to day 8, exchange cell culture medium with hiHep cell culture medium with 1% FBS every 2 days. Epithelial colonies usually form between day 5 and day 7.
10. On day 10, remove cell culture medium, and wash twice with 2 mL of PBS.
11. Trypsinize the cells with 1 mL of 0.05% trypsin at 37 °C for 5 min.
12. Suspend the cells with hiHep cell culture medium containing 1% FBS.
13. Transfer 5×10^5 cells to each 60-mm collagen I-coated dish. Add hiHep cell culture medium containing 1% FBS to 5 mL.

3.6 Enrichment of $hiHep^{LT}$ Cells

1. On day 12, exchange the hiHep cell culture medium with 1% FBS.
2. On day 14, aspirate the medium, and wash twice with 2 mL of PBS.
3. Add 1 mL of 2-mg/mL collagenase IV solution.
4. Incubate the cells for 5–10 min.
5. Carefully monitor the cells under a microscope. When nearly 50% of the cells curl up and start to detach from the dish surface, gently tap the dish until most of the fibroblasts dissociate, while most of epithelial colonies remain on the dish.
6. Aspirate the supernatant. Wash with 4 mL of PBS to remove collagenase IV.
7. Add 1 mL of 0.05% trypsin and incubate the cells at 37 °C for 3–5 min.
8. Suspend the cells with hiHep cell culture medium containing 1% FBS.

9. Transfer 5×10^5 cells to each 60-mm collagen I-coated dish. Add 5 mL of hiHep cell culture medium with 1% FBS.
10. Exchange cell culture medium with fresh hiHep cell culture medium with 1% FBS every 2 days until the cells become confluent (*see* **Note 4**).

3.7 Passage of HiHep^{LT} Cells

1. When the cells become confluent, discard the medium, and wash twice with 2 mL of PBS.
2. Trypsinize the cells with 1 mL of 0.05% trypsin at 37 °C for 5 min.
3. Add 2 mL of hiHep cell culture medium supplemented with 1% FBS.
4. Suspend the cells, and transfer 5×10^5 cells to each 60-mm collagen I-coated dish.
5. Add 5 mL of hiHep cell culture medium with 1% FBS.
6. Exchange medium with fresh hiHep cell culture medium without FBS every day.
7. For functional assays, usually use hiHep^{LT} cells 4–8 days after seeding (*see* **Note 5**).

4 Notes

1. The DMEM/F12 1:1 mixture already contains 0.5-mM sodium pyruvate. Add extra sodium pyruvate to 1 mM in the final human fibroblast culture medium.
2. The tissue pieces should be attached to the surface of the dish after adding HFC medium.
3. MOI and starting cell number significantly affect the efficiency of transdifferentiation. The best MOI and starting cell number vary for different HSF^{LT} strains. It is highly recommended to perform pre-experiments to test the best MOI and starting cell number for each HSF^{LT} strain.
4. We usually perform 1–2 rounds of enrichment to remove most of the cells that fail to be transdifferentiated.
5. HiHep^{LT} cells will undergo further maturation in FBS-free hiHep cell culture medium after confluence.

References

1. Scarpelli DG, Rao MS (1981) Differentiation of regenerating pancreatic cells into hepatocyte-like cells. Proc Natl Acad Sci U S A 78(4):2577–2581
2. Shen CN, Horb ME, Slack JM, Tosh D (2003) Transdifferentiation of pancreas to liver. Mech Dev 120(1):107–116
3. Reddy JK, Rao MS, Qureshi SA, Reddy MK, Scarpelli DG, Lalwani ND (1984) Induction

and origin of hepatocytes in rat pancreas. J Cell Biol 98(6):2082–2090

4. Shen CN, Slack JM, Tosh D (2000) Molecular basis of transdifferentiation of pancreas to liver. Nat Cell Biol 2(12):879–887. https://doi.org/10.1038/35046522
5. Huang P, He Z, Ji S, Sun H, Xiang D, Liu C, Hu Y, Wang X, Hui L (2011) Induction of functional hepatocyte-like cells from mouse fibroblasts by defined factors. Nature 475 (7356):386–389. https://doi.org/10.1038/nature10116
6. Huang P, Zhang L, Gao Y, He Z, Yao D, Wu Z, Cen J, Chen X, Liu C, Hu Y, Lai D, Hu Z, Chen L, Zhang Y, Cheng X, Ma X, Pan G, Wang X, Hui L (2014) Direct reprogramming of human fibroblasts to functional and expandable hepatocytes. Cell Stem Cell 14 (3):370–384. https://doi.org/10.1016/j.stem.2014.01.003
7. Shi XL, Gao YM, Yan YP, Ma HC, Sun LL, Huang PY, Ni X, Zhang LD, Zhao X, Ren HZ, Hu D, Zhou Y, Tian F, Ji Y, Cheng X, Pan GY, Ding YT, Hui LJ (2016) Improved survival of porcine acute liver failure by a bioartificial liver device implanted with induced human functional hepatocytes. Cell Res 26(2):206–216. https://doi.org/10.1038/cr.2016.6
8. Zhou T, Benda C, Dunzinger S, Huang YH, Ho JC, Yang JY, Wang Y, Zhang Y, Zhuang Q, Li YH, Bao XC, Tse HF, Grillari J, Grillari-Voglauer R, Pei DQ, Esteban MA (2012) Generation of human induced pluripotent stem cells from urine samples. Nat Protoc 7 (12):2080–2089. https://doi.org/10.1038/nprot.2012.115
9. Gao Y, Zhang X, Zhang L, Cen J, Ni X, Liao X, Yang C, Li Y, Chen X, Zhang Z, Shu Y, Cheng X, Hay DC, Lai D, Pan G, Wei G, Hui L (2017) Distinct gene expression and epigenetic signatures in hepatocyte-like cells produced by different strategies from the same donor. Stem Cell Reports 9(6):1813–1824. https://doi.org/10.1016/j.stemcr.2017.10.019
10. Du Y, Wang J, Jia J, Song N, Xiang C, Xu J, Hou Z, Su X, Liu B, Jiang T, Zhao D, Sun Y, Shu J, Guo Q, Yin M, Sun D, Lu S, Shi Y, Deng H (2014) Human hepatocytes with drug metabolic function induced from fibroblasts by lineage reprogramming. Cell Stem Cell 14 (3):394–403. https://doi.org/10.1016/j.stem.2014.01.008
11. Ji S, Zhu L, Gao Y, Zhang X, Yan Y, Cen J, Li R, Zeng R, Liao L, Hou C, Gao Y, Gao S, Wei G, Hui L (2017) Baf60b-mediated ATM-p53 activation blocks cell identity conversion by sensing chromatin opening. Cell Res 27(5):642–656. https://doi.org/10.1038/cr.2017.36

Chapter 10

Conversion of Fibroblasts to Hepatocyte-Like Cells In Vivo

Guangqi Song, Qinggong Yuan, Zhen Dai, Hsin-Chieh Tsay, Xizhong Shen, Michael Ott, and Amar Deep Sharma

Abstract

In vivo conversion of fibroblasts into hepatocyte-like cells provides one potential approach for the treatment of liver fibrosis. In our previous study, we showed in vivo conversion of myofibroblasts into induced hepatocytes (iHeps) by forced expression of four transcription factors in genetic fate-tracing mouse model of chronic liver disease. These in vivo-generated iHeps showed similar expression profile with endogenous hepatocytes (eHeps) and also exhibited similar functional characteristics, such as albumin secretion, urea synthesis, cytochrome activity, and drug responsiveness. Furthermore, the targeted expression of our reprogramming factors in myofibroblasts attenuated liver fibrosis. Our study suggests that in vivo reprogramming may open new perspectives for the treatment of diseases such as liver fibrosis.

Key words In vivo reprogramming, Induced hepatocytes (iHeps)

1 Introduction

Myofibroblasts contribute to liver fibrosis by secreting extracellular matrix in most types of chronic liver diseases. Therefore, we proposed an approach to ameliorate liver fibrosis by converting myofibroblasts directly into hepatocytes by introducing transcription factors. Recently, a novel strategy to convert fibroblasts into hepatocyte-like cells in vitro by forced expression of four transcription factors has been established [1–4]. In our recent study, we reported effectively trans-reprogramming of myofibroblasts derived from hepatic stellate cell (HSC) into hepatocyte-like cells (iHeps) by four transcription factors FOXA3, GATA4, HNF1A, and HNF4A (4TFs) in vitro. Furthermore, we showed in vivo converted myofibroblasts into iHeps by forced expression of 4TFs in genetic fate-tracing mouse model of chronic liver disease [5]. 4TFs were directly delivered into myofibroblasts using a p75NTRp-tagged recombinant adenoviral vector (serotype 5), which was designed to target mouse myofibroblasts through coupling with a peptide (S11-NGFp) [6]. S11-NGFp consisting of a

Naoki Tanimizu (ed.), *Hepatic Stem Cells: Methods and Protocols*, Methods in Molecular Biology, vol. 1905,
https://doi.org/10.1007/978-1-4939-8961-4_10, © Springer Science+Business Media, LLC, part of Springer Nature 2019

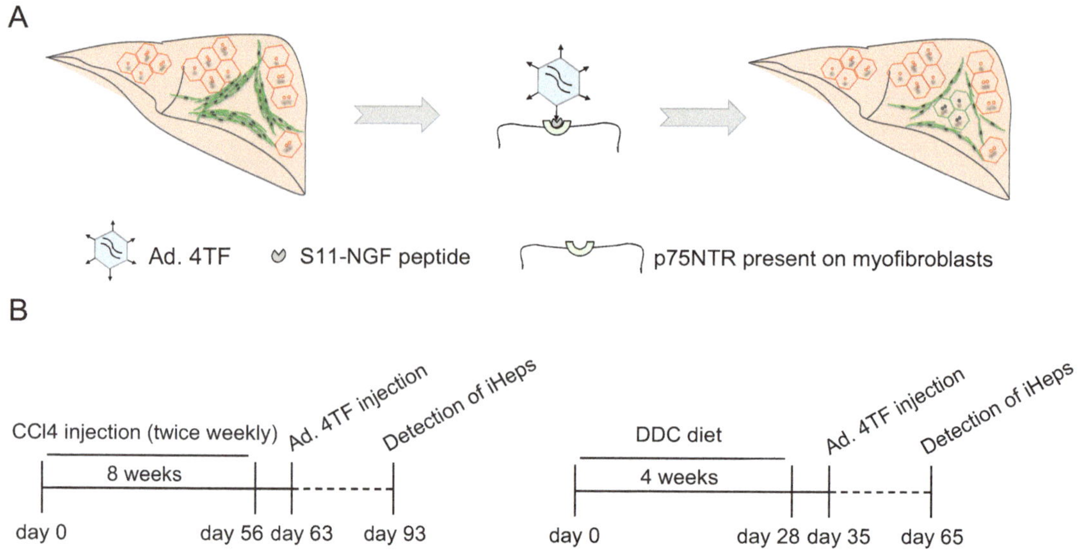

Fig. 1 (**a**) Schematic of in vivo conversion of myofibroblasts into iHeps in LratCre-mT/mG mice using p75NTRp-tagged recombinant adenoviral vector. In this lineage-tracing model, the endogenous hepatocytes (eHeps) would express membranous tdTomato, while myofibroblasts and induced hepatocytes (iHeps) derived from myofibroblasts would express membranous EGFP. (**b**) Schematic of in vivo iHep generation in CCl_4 and DDC diet-induced liver fibrosis disease mouse model

single chain antibody fragment (S11) directed against the Ad fiber knob and a nerve growth factor peptide (NGFp) that recognizes p75NTR on HSCs [6]. The in vivo-generated iHeps were observed in the liver of the lineage-tracing mice on the 30th day after viral injection. We sorted these in vivo-generated iHeps and characterize their function in vitro. They showed similar expression profile to endogenous hepatocytes (eHeps) and also exhibited similar functional characteristics, such as albumin secretion, urea synthesis, cytochrome activity, and drug responsiveness. Furthermore, conversion of myofibroblasts into iHeps in vivo ameliorated liver fibrosis. Both CCl_4 and DDC diet-induced fibrotic model showed lower level of liver fibrosis after treatment with adenovirus vector containing 4TFs. Our study suggested that in vivo trans-reprogramming may open new perspectives for the treatment of chronic liver diseases.

In this chapter, we describe our method for conversion of fibroblasts to iHeps in vivo (Fig. 1). Our strategy requires a lineage-tracing model and a modified target adenoviral vector system. LratCre-mT/mG mice have been shown to effectively trace myofibroblasts derived from the HSC lineage [7]. The modified adenoviral vector coupling with a peptide system has been shown to effectively target myofibroblasts in vivo [6]. Here, we describe a generation of iHeps using LratCre-mT/mG mice and p75NTRp-tagged adenoviral vector in vivo. Additionally, we describe characterization of in vivo-generated iHeps and evaluation of liver fibrosis level after in vivo conversion.

2 Materials

2.1 Animal Model

1. Gt(ROSA)26Sortm4(ACTB-tdTomato,-EGFP) mice (referred to as mT/mG).
2. Lrat-Cre mice.
3. 10% carbon tetrachloride (CCl_4) dissolved in corn oil.
4. 3,5-Diethoxycarbonyl-1,4-dihydrocollidine (0.1% DDC)-diet.
5. 3 M Vetbond™ No. 1469SB.

2.2 Cell and Molecular Biology

1. Ad.4TF: The human cDNAs of *FOXA3*, *GATA4*, *HNF1A*, and *HNF4A* were cloned together into a p75NTRp-tagged recombinant adenoviral vector (serotype 5, referred to as Ad.4TF).
2. A peptide (single chain antibody fragment) of the nerve growth factor (S11-NGFp), which was designed for coupling of Ad.4TF fiber knobs and binding to the p75 neurotrophin receptor (p75NTR) present on hepatic stellate cells and myofibroblasts.
3. pAdEasy1 plasmid.
4. *E. coli* BJ5183 bacterial cells.
5. HEK293 cells.
6. Caesium chloride (CsCl).
7. Ketamin.
8. Rompun (Bayer).
9. Perfusion solution: 0.5-mM ethylene glycol tetra acetic acid (EGTA) and 10-mM HEPES in Earle's Balanced Salt Solution (EBSS), pH 7.2–7.4.
10. Liberase solution: 100-μg/mL Liberase (Roche Liberase™ TL Research Grade) and 10-mM HEPES in EBSS, pH 7.2–7.4.
11. Dulbecco's Modified Eagle's Medium (DMEM).
12. Fetal calf serum (FCS) (PAN biotech).
13. 100-μm nylon filter.
14. Percoll.
15. Trypan blue.
16. Sorting buffer: 0.5% BSA and 1-mM EDTA in Ca/Mg-free PBS.
17. Collagen from rat tail tendon (Roche).

18. Hepatocyte Culture Medium BulletKit (HCM, Lonza), including 500-mL hepatocyte basal medium (HBM) and HCM SingleQuots Kit (contains ascorbic acid, 0.5 mL; BSA-FAF, 10 mL; hydrocortisone, 0.5 mL; human hEGF, 0.5 mL; transferrin, 0.5 mL; insulin, 0.5 mL; GA, 0.5 mL).
19. RNeasy Mini Kit (Qiagen).
20. TRIzol reagent (Invitrogen).
21. Whole Mouse Genome Oligo Microarray v2 (4 × 44K) (Agilent Technologies).
22. Albumin ELISA Quantitation Set (Bethyl Laboratories).
23. QuantiChrom Urea Assay Kit (BioAssay Systems).
24. P450-Glo CYP1A2, CYP2C9, and CYP3A assay kits (Promega).
25. Working solution of Oil red O: 90-mg Oil red O in 50-mL 60% isopropanol.
26. 1% Perid acid.
27. Schiff's reagent.
28. Methanol/acetone-mix (1:1): The solution should be keep at −20 °C.
29. DiI-ac-LDL.
30. Indocyanine green (ICG).
31. Glucagon (Sigma).
32. pCPT-cAMP (Sigma).
33. Insulin.
34. Glucose assay kit (Invitrogen).
35. Phenobarbital.
36. Rifampicin.
37. Agilent Genome Microarray Kits 4 × 180k arrays (G4125A, Agilent Technologies).
38. Primers and probes for quantitative real-time PCR (qPCR) (Tables 1 and 2).

2.3 Histology, Immunohistochemistry, and Immunofluorescence

1. Picro-Sirius Red solution: 0.1% direct red 80 plus 0.1% fast green dissolved in 1.2% saturated aqueous picric acid solution.
2. Antibodies for immunofluorescence staining: Albumin (Abcam-19196), MUP (Santa Cruz-21856), FAH (Abcam-81087), AAT (Abcam-117307), p75-NTR (Abcam-8874), desmin (Thermo Scientific-RB-9140), HNF4A (Santa Cruz-6556).
3. Hydroxyproline Assay Kit (Sigma).

Table 1
Primers and probes for detection of exogenous expression

Primers		
Gene	**Forward**	**Reverse**
HNF4A	CATGTACAGCTGCCGGTTCA	ACTGGTTCCGCTTATCCTTGTC
Taqman probe	**Catalog number**	
GATA4	Hs00171403_m1	Thermo Fisher
HNF1A	Hs01551752_m1	Thermo Fisher
FOXA3	Hs00270130_m1	Thermo Fisher

Table 2
Primers and probes for drug responsiveness assay

Primers		
Gene	**Forward**	**Reverse**
Cyp1a1	CCTCATGTACCTGGTAACCA	AAGGATGAATGCCGGAAGGT
Utg1a1	GGAGGCTGTTAGTGTTCCCT	CCGTCCAAGTTCCACCAAAG
Oatp	ATGGGATTCCATTCACTGGTTGTA	CACGTGCTCCACAGCTGGTTA
Taqman probe	**Catalog number**	
Abcc2	Mm00496899_m1	Thermo Fisher

3 Method

3.1 Production of Lineage-Tracing Model

1. For tracing in vivo conversion, a lineage-tracing model (LratCre-mT/mG mice) should be generated by crossing 12-week-old Lrat-Cre mice with mT/mG mice. Cells in LratCre-mT/mG mice express membrane-targeted tdTomato (mT) before Cre-mediated recombination. The expression of Cre is promoted by lecithin-retinol acyltransferase (Lrat) promoter, a hepatic stellate cells' specific promoter. LratCre-mT/mG mice have been shown as an effectively lineage-tracing model for myofibroblasts derived from the HSC lineage [7] (Fig. 2).
2. For inducing liver fibrosis, LratCre-mT/mG mice are injected with 4 μL/g of 10% CCl_4 dissolved in corn oil by intraperitoneal injection twice per week for 8 weeks (*see* **Note 1**).
3. For inducing cholestasis-induced liver fibrosis, LratCre-mT/mG mice are fed with DDC diet for 4 weeks.

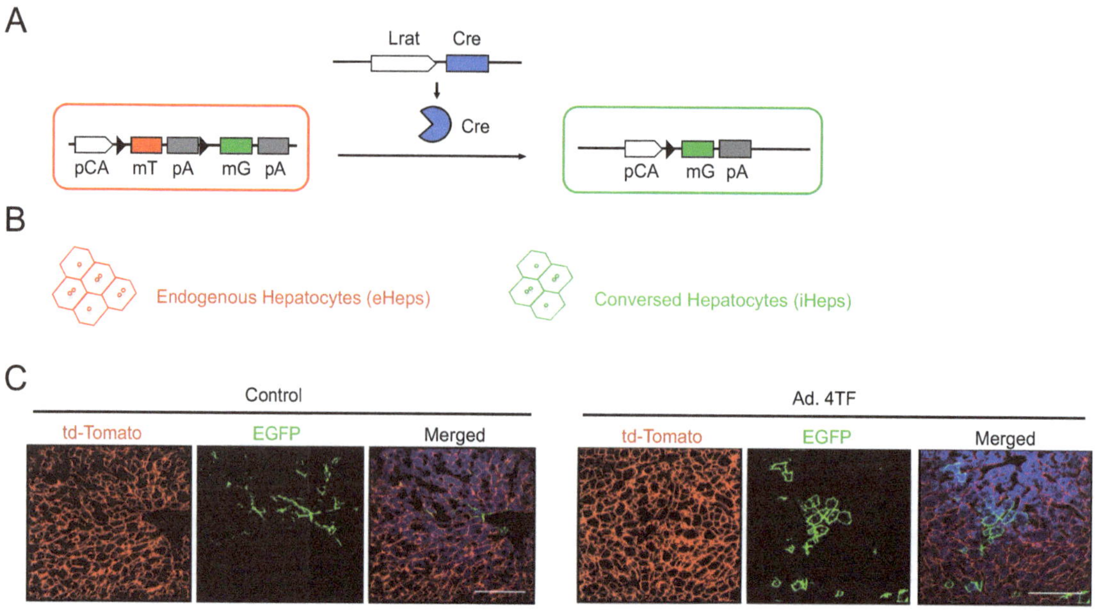

Fig. 2 (**a**) Cells in LratCre-mT/mG mice express membrane-targeted tdTomato (mT) before Cre-mediated recombination. Cre expressed under the transcriptional control of the lecithin-retinol acyltransferase (Lrat) promoter, which is specifically expressed in hepatic stellate cells in liver. Myofibroblasts derived from hepatic stellate cells express membrane-targeted EGFP. (**b**) Schematic illustration shows that eHeps can be identified by tdTomato expression, whereas iHeps are identified by EGFP expression. (**c**) These confocal figures show the generation of in vivo iHeps in LratCre-mT/mG mice after injected with Ad.4TF, no reprogrammed cell was found in control group

3.2 Conversion of Myofibroblasts into iHeps In Vivo

1. Ad.4TF expressing the *FOXA3*, *GATA4*, *HNF1A*, and *HNF4A* is generated by homologous recombination following co-transfection with pAdEasy1 in *E. coli* BJ5183. Ad.4TF should be propagated in HEK293 cells, purified by CsCl buoyant density centrifugation, and measured at OD260 [6]. The production and titration of the Ad.4TF vectors were performed following the method as previously published [8].
2. The peptide (S11-NGFp) (1 × 10^8 pfu virus with 100-μg peptide) should be preincubated with Ad.4TF for 1 h at 37 °C in serum-free DMEM before injection into LratCre-mT/mG mice.
3. One week after cessation of the CCl_4 or DDC diet treatment, p75NTRp-tagged Ad.4TF (Ad.4TF-S11-NGFp) should be injected into fibrotic LratCre-mT/mG mice. Briefly, mice were anesthetized using Ketamin and Rompun. The body cavity was opened, and 1 × 10^{10} Ad.4TF-S11-NGFp was injected via the portal vein using 30G needle (0.3 × 8 mm). After injection, press the whole of the portal vein for 1 min, and then add one drop of tissue adhesive (3 M Vetbond™

No. 1469SB) on the surface of the wound. The control group is injected with empty adenoviral vector.

4. 30 days after injection, liver samples should be collected from LratCre-mT/mG mice injected with Ad.4TF-S11-NGFp and empty adenoviral vector for producing frozen sections.
5. Confirmation of in vivo iHep conversed from myofibroblasts under fluorescence microscope. The eHeps would express membranous tdTomato, while the iHeps could be detected by membranous EGFP (Fig. 2).
6. Performing immunofluorescence staining for albumin, MUP, FAH, AAT, and HNF4A, to further confirm the in vivo conversion of myofibroblasts into iHeps expressing conventional hepatocyte markers.

3.3 Isolation of eHeps and iHeps

1. Hepatocytes were isolated from mice by a modified two-step Liberase perfusion [9]. Briefly, mice were anesthetized using Ketamin and Rompun. The body cavity was opened, and a catheter was placed into the portal vein and connected to a flow pump, which pumped media pre-warmed at 37 °C from a water bath into the catheter, and vena cava was cut. The liver was first perfused with perfusion solution. Subsequently, Liberase solution was applied for enzymatic digestion of the tissue at 37 °C. After digestion of 10–12 min, the liver was carefully disconnected, and the tissue was manually disrupted with sterile scissors and a scalpel in DMEM containing 10% FCS. The suspended hepatocytes were passed through a 100-μm nylon filter into 50-mL Falcon tubes. The cell suspensions were centrifuged twice at 300 rpm (or 50 × *g*) for 5 min at 4 °C, and the cell pellet was re-suspended in ice-cold DMEM medium containing 10% FCS. Cell viability was tested by trypan blue.
2. Centrifuge the cells for 5 min 300 rpm (or 50 × *g*), 4 °C, and re-suspend resulting pellet in sorting buffer. The concentration should be in the range from 0.5×10^6 to 5×10^6 cells/mL. Perform fluorescence-activated cell sorting (FACS) to separate eHeps and iHeps (*see* **Note 2**). Collect cells into DMEM medium containing 10% FCS; tdTomato-positive cells are eHeps, while EGFP-positive cells are iHeps (Fig. 3, *see* **Note 3**).
3. Preparation of the collagen-coated plate. Dissolve collagen with 0.2% acetic acid to a final concentration of 2 mg/mL. Add the solution into a plate following 2.5 μL per 1 cm^2, and cover the surface using cell scraper. It is better to stand the plate coated with collagen in biological safety cabinet more than 15 min to dry them enough before use.
4. The eHeps and iHeps should be seeded in collagen-coated plates, cultured in DMEM medium containing 10% FCS,

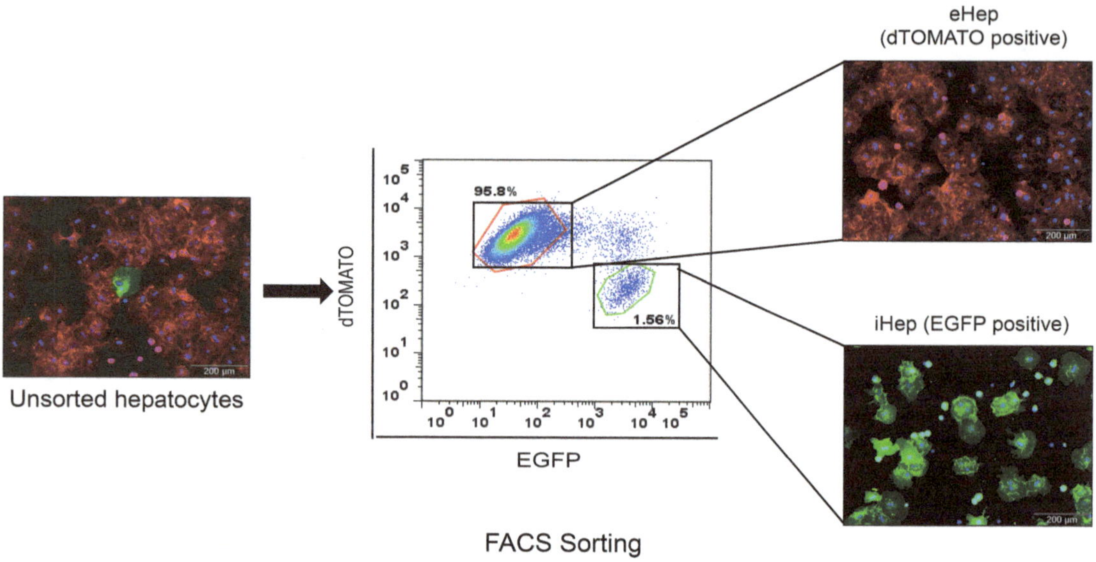

Fig. 3 Isolation of in vivo-generated iHeps (EGFP positive) and eHeps (tdTomato positive) by FACS sorting

37 °C, 5% CO_2. After 3 h, wash cells with PBS and replace the medium by HCM. To maintain hepatocyte, HCM was prepared by adding all of the reagents from SingleQuots Kit into HBM medium. The medium should be changed every other day.

5. Isolation of mRNA and functional assays should be performed after 24 h in culture with HCM. Those eHeps isolated from the same mouse can be used as positive control for iHeps in all characterization assays. Primary mouse hepatic stellate cells can be isolated from BALB/c mice according to previously published reports [10, 11]. These cells could be cultured in presence of PDGF to generate myofibroblasts, which can be used as negative control for iHeps.

3.4 Characterization of In Vivo Generated iHeps

1. For analysis of the stability of conversion, total RNA of iHeps should be isolated. Real-time qPCR can be performed using specific primers for four human factors to confirm exogenous silence. Array-CGH can be performed to test the genomic integrity of reprogrammed cells using Agilent Genome Microarray Kits 4 × 180k arrays with median overall probe spacing of about 13 kb. Labeling and hybridization of gDNA should be performed according to the protocol provided by Agilent.
2. For analysis of the global gene expression profiles, total RNA of eHeps, iHeps, and myofibroblasts should be isolated from cells by TRIzol reagent following a standardized protocol. Whole Mouse Genome Oligo Microarray v2 (4 × 44K) (Agilent Technologies) could be used to characterize global gene

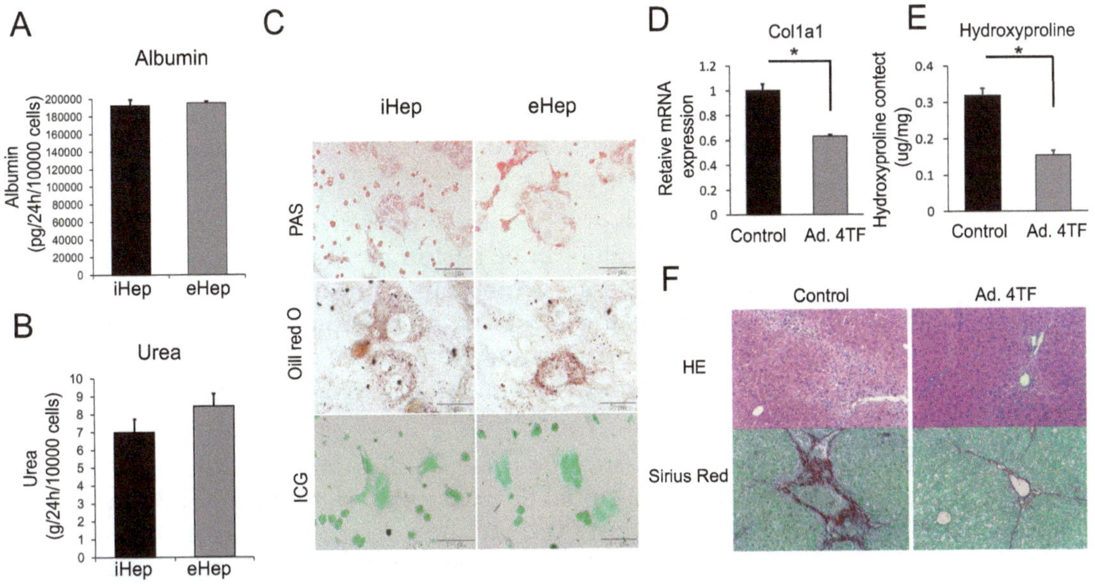

Fig. 4 (**a**) Albumin ELISA revealed comparable levels of secreted albumin in iHeps and eHeps. (**b**) Urea synthesis showed similar levels in iHeps and eHeps. (**c**) The iHeps show PAS staining, Oil red O staining, and ICG uptake. (**d–f**) Amelioration of liver fibrosis in Ad.4TF-injected mice. (**d**) Reduced levels of Col1a1 mRNA in Ad.4TF-injected mice compared to control mice. (**e**) Hydroxyproline assay showed decreased levels of entire collagen content, measured in whole liver. (**f**) H&E and Sirius Red staining showed less fibrosis in Ad.4TF-injected mice than respective controls

expression profiles of iHeps compared to myofibroblasts and eHeps.

3. For analysis of albumin secretion, eHeps and iHeps were washed with PBS and cultivated in HCM without BSA-FAF from SingleQuots Kit. After 24 h, collect supernatants, and measure albumin secretion using the mouse albumin ELISA Quantitation Set according to the manufacturer's instructions. Results should be normalized to 10,000 cells at 24 h (Fig. 4a).
4. For analysis of urea production, wash eHeps and iHeps with PBS, and replace the medium with HCM without BSA-FAF from SingleQuots Kit. After 24 h, collect supernatants, and measure urea production using the QuantiChrom Urea Assay Kit according to the manufacturer's instructions. Results should be normalized to 10,000 cells at 24 h (Fig. 4b).
5. For analysis of cytochrome P450 enzymes activities, P450-Glo CYP1A1, CYP1A2, CYP2C9, and CYP3A assay kits (Promega) could be used to measure CYP1A1, CYP1A2, CYP2C9, and CYP3A activities in eHeps and iHeps. Seed cells in collagen-coated 24 well plates, 2–3 × 10^4 per well. Add specific inducers into medium, incubate for 48–72 h according to manufacturer's instructions (cell-based, non-lytic protocol), and change medium everyday with inducers. Wash cell using PBS

and replace by HCM medium without inducers. Add specific substrates for CYP1A1, CYP1A2, CYP2C9, and CYP3A into medium separately. After incubation, perform measurements according to manufacturer's instructions. The luminescence signals from the P450-Glo assay were normalized to cell numbers (*see* **Note 4**).

6. For analysis of the ability to store glycogen, periodic acid-Schiff (PAS) staining can be performed. Fix eHeps and iHeps with a cold methanol/acetone-mix for 10 min at 4 °C. Add 1% periodic acid, and incubate the at room temperature for 5 min. Wash cells with water, and incubate cells with Schiff's reagents for 15 min; examine the store glycogen under a phase contrast microscope (Fig. 4c).
7. For analysis of the presence of triglycerides and lipids, eHeps and iHeps should be fixed using 4% formalin for 30 min. Wash one time with PBS, add the working solution of Oil red O, and incubate for 10 min. Remove the working solution and wash two times with 60% isopropanol. Add the hematoxylin counterstain and incubate for 1 min, washed with PBS and examined under a phase contrast microscope (Fig. 4c, *see* **Note 5**).
8. For analysis of the ability to uptake low-density lipoprotein (LDL), eHeps and iHeps should be incubated at 37 °C for 4 h with 10-μg/mL DiI-Ac-LDL, washed with PBS and observed by fluorescence microscopy.
9. For analysis of the ability to uptake indocyanine green (ICG), eHeps and iHeps should be incubated at 37 °C for 3 h with 0.1-mg/mL ICG, washed with PBS and examined under a phase contrast microscope (Fig. 4c).
10. For analysis of the glucose production, eHeps and iHeps should be treated with either glucagon or insulin. For glucagon treatment, eHeps and iHeps should be seeded on collagen-coated plates and cultured in HCM medium with or without glucagon (100 nM). Similarly, for insulin treatment, cells should be cultured in HCM medium supplemented with 0.5-mM pCPT-cAMP and with or without 100-nM insulin. After 2 h incubation, collect the supernatant, and analyze for glucose production using a glucose assay kit (Invitrogen).
11. For analysis of the drug responsiveness, eHeps and iHeps should be cultured in the HCM medium with or without 1-mM phenobarbital for 72 h. Total RNA should be isolated to perform real-time qPCR, and compare the expression level of Cyp1a1 and Abcc2. And eHeps and iHeps should be cultured in the HCM medium with or without 1-mM 50-μM Rifampicin for 72 h. Total RNA should be isolated to perform real-time qPCR, and compare the expression level of Ugt1a1 and Oatp.

3.5 Evaluation of Liver Fibrosis Level After In Vivo Conversion

1. For analysis of liver fibrosis level of mice injected with Ad.4TF and empty adenoviral vector, the expression level of Col1a1 in the entire liver could be detected by real-time qPCR. The total RNA should be isolated from liver tissues that were collected on the 30th day after viral injection (Fig. 4d).
2. The entire collagen content of a liver could be detected by hydroxyproline assay. Liver tissues should be collected from mice injected with Ad.4TF and empty adenoviral vector, and perform assay following the manual of Hydroxyproline Assay Kit (Fig. 4e).
3. Paraffine sections of liver tissues from mice injected with Ad.4TF and empty adenoviral vector can be produced following a standardized protocol. For analysis of the fibrotic area, Sirius Red staining could be performed. Following deparaffinization, the sections should be stained with Picro-Sirius Red solution and incubated for 60 min. Sections were rinsed with water, dehydrated and mounted in xylene. Take ten random photos from each section under a phase contrast microscope. Fibrotic area could be counted using Image J software (Fig. 4f).
4. For counting the number of fibroblasts in liver tissues, Desmin and p75-NTR staining can be performed following a standardized immunohistochemistry staining protocol.

4 Notes

1. When you generate liver fibrosis using 10% CCl_4, mice should be monitored carefully. It is highly possible to result in mice's death due to overdose injection. You can inject 10% CCl_4 two times per week as shown in Table 3.
2. Conversed iHeps show similar volume to hepatocytes; you should use 100-μm nylon filter after isolating them from livers. And iHeps can fall down to the bottom of Falcon tubes in 10 min. Thus, when you perform FACS sorting, invert the tube gently every 10 min, and use 100-μm nozzle to prevent cells from blocking the machine.
3. We performed FACS sorting; tdTomato and GFP double-positive cells could be found in the cells isolated from lineage-tracing mice. We confirmed that all hepatocytes (and rare double-positive cells) isolated from control mice are tdTomato positive. Since LratCre-mTmG model allows specific labeling of hepatic stellate cells and myofibroblasts with GFP, we believe that the double-positive cells in a LratCre-mT/mG model may represent doublets of hepatocytes and myofibroblasts.
4. For the cytochrome P450 enzymes' activities assay, the medium should be change every day with inducers. For

Table 3
Dose of CCl_4

Mouse weight (g)	10% CCl_4 vol. (μL)	Mouse weight (g)	10% CCl_4 vol. (μL)	Mouse weight (g)	10% CCl_4 vol. (μL)
14	56	23	92	32	128
15	60	24	96	33	132
16	64	25	100	34	136
17	68	26	104	35	140
18	72	27	108	36	144
19	76	28	112	37	148
20	80	29	116	38	152
21	84	30	120	39	156
22	88	31	124	40	160

CYP1A2/Luciferin-1A2 assay, replace HCM medium by Krebs-Henseleit buffer for incubation of substrate. And, it is better to perform the measurement immediately after incubation with substrate. Do not store the supernatant in −20 °C or −80 °C before measurement.

5. Oil red O working solution is stable for no longer than 2 h. You can prepare the stock solution by dissolving 300 mg of Oil red O powder in 100-mL 99% isopropanol. This stock solution is stable for 1 year in the dark with 4 °C temperature. Prepare work solution by mixing 30 mL of stock solution with 20-mL deionized water. Filter the working solution completely through a filter funnel before use.

References

1. Huang PY, He ZY, Ji SY, Sun HW, Xiang D, Liu CC, Hu YP, Wang X, Hui LJ (2011) Induction of functional hepatocyte-like cells from mouse fibroblasts by defined factors. Nature 475:386–U142
2. Huang P, Zhang L, Gao Y, He Z, Yao D, Wu Z, Cen J, Chen X, Liu C, Hu Y et al (2014) Direct reprogramming of human fibroblasts to functional and expandable hepatocytes. Cell Stem Cell 14:370–384
3. Sekiya S, Suzuki A (2011) Direct conversion of mouse fibroblasts to hepatocyte-like cells by defined factors. Nature 475:390–393
4. Yang R, Zheng Y, Li L, Liu S, Burrows M, Wei Z, Nace A, Herlyn M, Cui R, Guo W et al (2014) Direct conversion of mouse and human fibroblasts to functional melanocytes by defined factors. Nat Commun 5:5807
5. Song G, Pacher M, Balakrishnan A, Yuan Q, Tsay HC, Yang D, Reetz J, Brandes S, Dai Z, Putzer BM et al (2016) Direct reprogramming of hepatic myofibroblasts into hepatocytes in vivo attenuates liver fibrosis. Cell Stem Cell 18:797–808
6. Reetz J, Genz B, Meier C, Kowtharapu BS, Timm F, Vollmar B, Herchenroder O, Abshagen K, Putzer BM (2013) Development of adenoviral delivery systems to target hepatic stellate cells in vivo. PLoS One 8:e67091
7. Mederacke I, Hsu CC, Troeger JS, Huebener P, Mu X, Dapito DH, Pradere JP, Schwabe RF (2013) Fate tracing reveals hepatic

stellate cells as dominant contributors to liver fibrosis independent of its aetiology. Nat Commun 4:2823

8. Armendariz-Borunda J, Bastidas-Ramirez BE, Sandoval-Rodriguez A, Gonzalez-Cuevas J, Gomez-Meda B, Garcia-Banuelos J (2011) Production of first generation adenoviral vectors for preclinical protocols: amplification, purification and functional titration. J Biosci Bioeng 112:415–421
9. Yuan Q, Loya K, Rani B, Mobus S, Balakrishnan A, Lamle J, Cathomen T, Vogel A, Manns MP, Ott M et al (2013) MicroRNA-221 overexpression accelerates hepatocyte proliferation during liver regeneration. Hepatology 57:299–310
10. Bartneck M, Warzecha KT, Tag CG, Sauer-Lehnen S, Heymann F, Trautwein C, Weiskirchen R, Tacke F (2015) Isolation and time lapse microscopy of highly pure hepatic stellate cells. Anal Cell Pathol (Amst) 2015:417023
11. Weiskirchen R, Gressner AM (2005) Isolation and culture of hepatic stellate cells. Methods Mol Med 117:99–113

Chapter 11

Chemically Induced Liver Progenitors (CLiPs): A Novel Cell Source for Hepatocytes and Biliary Epithelial Cells

Takeshi Katsuda and Takahiro Ochiya

Abstract

Bipotent liver progenitor cells (LPCs) are promising cell sources for cell transplantation therapy in hepatic disorders as well as biliary dysfunctions. Using a cocktail of small molecules, we recently reported a novel approach to generate bipotent LPCs, named chemically induced liver progenitors (CLiPs), from adult rat hepatocytes. In this chapter, we describe a detailed protocol for the induction of rat CLiPs. We first describe the method to isolate primary rat hepatocytes and then describe how to induce CLiPs from the hepatocytes. In addition, we describe methods to induce the generated CLiPs to differentiate into hepatocytes and biliary epithelial cells.

Key words Hepatocyte, Biliary epithelial cell, Liver progenitor cell, CLiP, Reprogramming, Bipotentiality

1 Introduction

The only current treatment for end-stage liver diseases is liver transplantation, but its application is limited due to donor shortages. Thus, there is a strong demand for a novel cell source to realize regenerative medicine for liver diseases. Accordingly, researchers have proposed various approaches to supply potential hepatic cell sources, using embryonic stem cells [1], fetal liver cells [2], induced pluripotent stem cells [3, 4], iHep cells [5, 6], and adult liver progenitor cells (LPCs) [7, 8]. These advances have attracted much attention, but there are still hurdles to be overcome, such as ethical issues involved with fetal tissue, relatively immature functionality of the generated hepatic cells compared to mature hepatocytes (MHs), and poor availability of the adult progenitor cells. Therefore, further exploration for novel hepatic cell sources is required to realize liver regenerative medicine.

We recently proposed a novel type of LPCs that can stably expand in vitro and exhibit bipotentiality [9]. Using a cocktail of three small molecules Y-27632, A-83-01, and CHIR99021

Naoki Tanimizu (ed.), *Hepatic Stem Cells: Methods and Protocols*, Methods in Molecular Biology, vol. 1905, https://doi.org/10.1007/978-1-4939-8961-4_11, © Springer Science+Business Media, LLC, part of Springer Nature 2019

(termed YAC), this approach allows reprogramming of rat MHs into bipotent LPCs without any genetic modification. Thus, we named these cells chemically induced liver progenitors (CLiPs). Rat CLiPs can efficiently repopulate injured liver of chronic hepatitis animal models. Moreover, transplanted CLiPs, at least partly, contribute to ductal structure formation, implying their potential applicability also to regeneration of the biliary system [9]. Here, we describe a detailed protocol to generate CLiPs from primary rat MHs and to evaluate the bipotentiality of the generated CLiPs.

2 Materials

2.1 Animal and Reagents

1. Wistar rats aged at 5 weeks to 20 weeks (both male and female animals can be used).
2. Pre-perfusion buffer: Dissolve 40 g NaCl, 2.0 g KCl, 0.39 g $NaH_2PO_4{\cdot}2H_2O$, 0.76 g $Na_2HPO_4{\cdot}12H_2O$, 4.5 g glucose, 1.9 g EGTA, 3.7 g EDTA, 0.0030 g phenol red, and 11.9 g HEPES in 500 mL distilled water to prepare 10× stock solution. Dilute 50 mL of 10× stock solution with 450 mL distilled water. After autoclave sterilization, add 2.35 mL of 7.5% $NaHCO_3$ solution and 2 mL of 1 N NaOH solution.
3. 7.5% $NaHCO_3$ solution: Dissolve 7.5 g of $NaHCO_3$ in 100 mL dH_2O. Filter with a 0.22 μm filter, and store at 4 °C until use. Use within 1 month.
4. 0.05% collagenase solution: Dissolve 9.6 g NaCl, 0.48 g KCl, 0.094 g $NaH_2PO4{\cdot}2H_2O$, 0.18 g $Na_2HPO_4{\cdot}12H_2O$, 2.86 g HEPES, 0.888 g $CaCl_2{\cdot}2H_2O$, and 0.42 g $NaHCO_3$ in 1.2 L of dH_2O supplemented with 14.4 mL of 0.05% phenol red solution. Store this basal solution at 4 °C until use. For perfusion, pour 400 mL into a beaker and adjust pH to 7–7.4 with 5 N NaOH and/or 6 N HCl. Then, dissolve 0.02 g trypsin inhibitor (Sigma-Aldrich) and 0.2 g collagenase (Wako). After adjusting the pH to 7.6 with 5 N NaOH and/or 6 N HCl, filter with a 0.22 μm filter, and store at 4 °C until use. The prepared collagenase solution should be used within 1 week.
5. Hepatocyte wash medium: E-MEM.
6. Percoll (GE Healthcare).
7. L-15-based medium for complete Percoll solution: Dissolve the following reagents in 99 mL of Leibovitz's L-15 medium (Gibco, with additives: Sodium pyruvate; L-Glutamine; Phenol red): 0.2 g bovine serum albumin (BSA) (Sigma-Aldrich) and 0.0429 g HEPES (Sigma-Aldrich). Filter with a 0.22 μm filter, add 1 mL of antibiotic/antimycotic (Gibco), and store at 4 °C until use.
8. 0.4% trypan blue solution.

9. Complete Percoll solution: Prepare complete Percoll solution in a 50 mL conical tube just before use by mixing 25 mL of L-15-based medium, 2.4 mL of 10× HBSS (Gibco), and 21.6 mL of Percoll solution.
10. 2 M HEPES solution: Dissolve 47.7 g HEPES in 100 mL dH_2O. Filter with a 0.22 μm filter, and store at 4 °C until use.
11. L-proline stock solution: Dissolve 0.3 g of L-proline (Sigma-Aldrich) in 10 mL of PBS(−). Filter with a 0.22 μm filter, and store at 4 °C until use.
12. 5% BSA stock solution: Dissolve 5 g BSA in 100 mL of PBS(−). Filter with a 0.22 μm filter. Protect from light and store at 4 °C until use.
13. 10 μg/mL EGF stock solution: Prepare 20 mL of 0.1% BSA solution by diluting 0.4 mL of 5% BSA with 19.6 mL PBS(−), and dissolve 0.2 mg EGF (Sigma-Aldrich). Store 500 μL aliquots at −20 °C until use.
14. 10^{-4} M dexamethasone: Dissolve 100 mg dexamethasone in 25.4 mL ethanol to make 10^{-2} M solution and store 500 μL aliquots at −20 °C until use. Dilute 500 μL of 10^{-2} M solution with 49.5 mL dH_2O. Store the stock solution at 4 °C in the dark.
15. 1 M nicotinamide: Dissolve 12.21 g nicotinamide (Sigma-Aldrich) in 100 mL PBS(−). Filter with a 0.22 μm filter, and store at 4 °C until use.
16. 100 mM Asc2P stock solution: Dissolve 2.9 g of Asc2P (Wako) in 100 mL PBS(−). Filter with a 0.22 μm filter. Protect from light and store at 4 °C until use.
17. 5 mM Y-27632: Dissolve 25 mg of Y-27632 (Wako) in 14.8 mL sterilized dH_2O. Store 1 mL aliquots at −20 °C until use.
18. 2.5 mM A-83-01: Dissolve 10 mg of A-83-01 (Wako) in 949 μL DMSO to make 25 mM solution and make 100 μL aliquots for long-term preservation at −20 °C. Add 900 μL DMSO to one vial of 25 mM solution, and store 100 μL aliquots at −20 °C.
19. 15 mM CHIR99021: Add 3.58 mL DMSO into the vial containing 25 mg CHIR99021 (Axon Medchem), and store 100 μL aliquots at −20 °C.
20. Small hepatocyte medium (SHM): Remove 27 mL from a 500 mL bottle of DMEM/F12 (with 2.4 g/L $NaHCO_3$ and L-glutamine, Life Technologies, #11320–082)). The removed DMEM/F12 can be used to prepare Matrigel-coated plates (*see* below). To the remaining DMEM/F12, add 1.25 mL of 2 M HEPES, 500 μL L-proline, and 5 mL antibiotic/

antimycotic. Then, add 250 μL of 5 N NaOH to adjust pH to approximately 7.5. Then, add 5 mL of 5% BSA, 500 μL of 10 μg/mL of EGF, 5 mL of ITS-X, 500 μL of 10^{-4} M dexamethasone, 5 mL of 1 M nicotinamide, and 5 mL of 100 mM Asc2P. Store at 4 °C.

21. 10 μg/mL mouse oncostatin M (OSM) stock solution: Prepare 2.5 mL of 0.1% BSA/PBS(−) by diluting 50 μL of 5% BSA with 2.45 mL PBS(−). Dissolve 25 μg mouse OsM (R and D Systems) in 2.5 mL of 0.1% BSA solution. Store 100 μL aliquots at −20 °C until use.
22. Matrigel-coated plate: Dilute Matrigel (Corning) at a ratio of 16 μL per mL of DMEM/F12. Add approximately 0.1–0.2 mL/cm^2 of the diluted Matrigel solution. Incubate at 37 °C for 20 min. Wash once with an equal volume of SHM (−) before seeding cells.
23. Hepatic induction medium for Step 1 (HIM-1): Add 100 μL of 10 μg/mL OsM and 500 μL of 10^{-4} M dexamethasone into 50 mL of SHM + YAC. Final concentrations of OsM and dexamethasone are 20 ng/mL and 10^{-6} M, respectively.
24. Hepatic induction medium for Step 2 (HIM-2): Prepare this medium on ice immediately before use. Mix HIM-1 and Matrigel at 7:1 volume ratio. For example, mix 62.5 μL Matrigel and 437.5 μL HIM-1 (500 μL in total) and 125 μL Matrigel and 875 μL HIM-1 (1 mL in total) for one well of a 24- or 12-well plate, respectively. Use P1000 tips or wide-bore P200 tips chilled at 4 °C or on ice when working with Matrigel.
25. DMEM, high glucose, pyruvate.
26. Cryopreserved MEFs (EmbryoMax® Primary Mouse Embryo Fibroblasts) (Millipore, catalog number: PMEF-CF).
27. BEC induction medium for Step 1 (BIM-1): BIM-1 is mTeSR™1 medium (STEMCELL Technologies, catalog number: ST-05850) supplemented with YAC. Prepare complete mTeSR™1 medium by adding the 100 mL of supplement into 400 mL basal medium, and add each vial of YAC stock to the solution.
28. BEC induction medium for Step 2 (BIM-2): BIM-2 is BIM-1 supplemented with 2% Matrigel. Prepare this medium immediately before use. Add 1 volume of Matrigel to 49 volumes of BIM-1.

2.2 Equipment

1. Peristaltic pump (EYELA, Tokyo, Japan, catalog number: RP-1000).
2. Silicon tube (Ø4.76 × 7.94 mm^2) (EYELA, catalog number: 125540).

3. Cannula Ø1.2 mm (plastic outer needle equipped with intravenous cannula) (Top corporation, Tokyo, Japan, catalog number: V1).
4. 4–0 sterilized braid silk suture (Akiyama Medical Mfg, Co. Ltd., catalog number: DEWB0404).
5. Stainless 60 μm cell strainer (Ikemoto Scientific Technology, Tokyo, Japan, catalog number: 802–570-01).

3 Methods

3.1 Isolation of Rat Primary MHs

Rat primary MHs are isolated using the two-step collagenase perfusion method [10]. In the first step, blood is eliminated from the liver by perfusing the liver with pre-perfusion buffer. In the second step, the liver is digested by perfusing collagenase solution. Then, the liver is extracted, minced, and further digested ex vivo with the remaining collagenase solution. MHs are collected by sequential low-speed centrifugation at 57 × *g* (*see* **Note 1**). This sequential low-speed centrifugation allows exclusion of nonparenchymal cells.

1. Before starting, place 20 mL E-MEM at room temperature, which will be used for additional digestion of the extracted liver.
2. Anesthetize a rat by inhalation of isoflurane vapor. Place a paper towel cushion on a foam board or cork board, and lay the rat on the cushion. Sterilize the animal by wiping its fur with 70% EtOH. Open the abdomen using surgical scissors to expose the liver.
3. Start drip of pre-perfusion buffer at 25–30 mL/min. Half-cut the portal vein at the location approximately 1.5 cm from the bifurcation of the portal vein, and insert the cannula.
4. Immediately after confirming that the color of the liver changes from reddish brown to yellowish, cut the inferior vena cava. Perfuse approximately 450 mL pre-perfusion buffer. It takes approximately 15 min.
5. Pause the pump, transfer the inlet tube to the collagenase bottle, and resume perfusion. Perfuse approximately 300–350 mL collagenase solution. It takes approximately 10 min.
6. Stop perfusion, and remove the cannula from the portal vein. Cut the liver from the abdominal cavity, and transfer it to a 100 mm sterile plastic dish. Remove the attaching surrounding tissue, such as the diaphragm (*see* **Note 2**).
7. Add 20 mL E-MEM kept at room temperature and 20 mL of the remaining collagenase solution. Mince the liver with

surgical scissors into pieces that are smaller than approximately 3 mm in size.

8. Incubate the minced tissue at 37 °C in a CO_2 incubator for 15 min. After taking out the dish from the incubator, further dissociate the remaining tissue by gently pipetting the digested solution with a wide-bore pipette, such as a 25 mL pipette, several times (*see* **Notes 3** and **4**).
9. Remove the undigested tissue fragments by transferring the digested solution into a 100 mL tube equipped with single-layered gauze on the top. Discard the single-layered gauze from the tube top, and transfer the filtered solution into a 100 mL tube with the double-layered gauze (*see* **Notes 5** and **6**).
10. Aliquot the filtered solution into two 50 mL tubes. Add E-MEM into each tube to 50 mL by decantation, and mix by gentle inversion. Centrifuge the tubes at 57 × *g* for 1 min at 4 °C.
11. Aspirate the supernatant, and loosen the cell pellets by gently tapping the tubes. Add 10 mL E-MEM to each tube, and resuspend the cell pellet by gently pipetting several times. Decant E-MEM into each tube to 50 mL, and mix by gentle inversion. Filter the cell suspension through a double-meshed stainless cell strainer into new tubes. Centrifuge the tubes at 57 × *g* for 1 min at 4 °C.
12. Aspirate the supernatant, and loosen the cell pellet by gently tapping the tubes. Add 10 mL complete Percoll solution to each tube, and resuspend the cell pellet by gently pipetting several times until the cells are suspended homogeneously. Aliquot the remaining complete Percoll solution to each tube (approximately 14 mL/tube), and mix by gentle inversion. Centrifuge the tubes at 57 × *g* for 10 min at 4 °C.
13. Aspirate the supernatant, and loosen the cell pellet by gently tapping the tubes. Add 10 mL E-MEM to each tube, and resuspend the cell pellet by gently pipetting several times. Decant E-MEM into each tube to 50 mL, and mix by gentle inversion. Centrifuge the tubes at 57 × *g* for 2 min at 4 °C.
14. Repeat the above wash process (**Step 13**) once more.
15. Resuspend the cells in 15 mL SHM(−), combine the two tubes.
16. (Optional) If extensive aggregation disturbs cell counting, filter the cell suspension with a 40 μm cell strainer before cell counting. There is no difference in reprogramming efficiency under microscopic observation between cells isolated with and without this filtration step.

17. Mix 30 μL of cell suspension with 120 μL SHM to prepare a fivefold-diluted cell suspension. Remove 30 μL from the diluted cell suspension, and mix it with 30 μL of trypan blue. Count the viable and dead cell numbers using a hemocytometer. Determine the suspension volume to be used for culture.
18. Seed the isolated MHs at approximately 1×10^5 cells/cm^2. To obtain homogeneous seeding of MHs, shake the plates vigorously back and forth ten times and then right and left ten times (*see* **Note 7**). Let the plates stand still for 15 min so that MHs fall down to the bottom of the plate, and then, transfer the plates into a CO_2 incubator.

3.2 Induction of Primary CLiPs Using the Small Molecule Cocktail YAC

Rat CLiPs can be induced from primary MHs upon stimulation with a combination of three small molecules, YAC. An evident morphological change compared with YAC-free (YAC(−)) culture occurs approximately 4–5 days after plating. As previously reported, rodent MHs can divide several times in vitro even without YAC, but their proliferation completely stops thereafter [11, 12]. In contrast, MHs can continuously proliferate to produce CLiPs during 2-week culture (Fig. 1). When CLiPs reach 70–100% confluency, harvest CLiPs to perform hepatic induction or biliary induction, establish stable CLiPs, or prepare frozen stocks for future use.

1. One day after plating primary MHs, replace the culture medium with fresh media. Thereafter, renew the medium every 2–3 days. Given that proliferative YAC(+) cells are bipotential (*see* the following sections), we designate these cells as chemically induced liver progenitors (CLiPs).
2. Wash plates with PBS(−) twice.
3. Add 0.05% trypsin-EDTA or TrypLE Express (these two reagents can be used equivalently for harvesting CLiPs), and incubate at 37 °C in a CO_2 incubator until the cells are detached from the plate (*see* **Notes 8** and **9**).
4. Further dissociate the cells physically by pipetting with a P1000 tip to harvest them as completely as possible, and transfer them to a 15 mL or 50 mL collection tube (*see* **Note 10**).
5. Collect the cells remaining in the plate with 5 mL SHM + YAC supplemented with 5% FBS or SHM + YAC for trypsin-EDTA or TrypLE Express (TrypLE Express does not require neutralization of the enzymatic activity).
6. Count the cell number using a hemocytometer, determine the suspension volume to be used for passage, transfer the required volume to a new tube, and centrifuge the tube at $200 \times g$ for 5 min. Aspirate the supernatant, and loosen the cell pellet by tapping the tubes. Resuspend the cells with medium to be used for desired application.

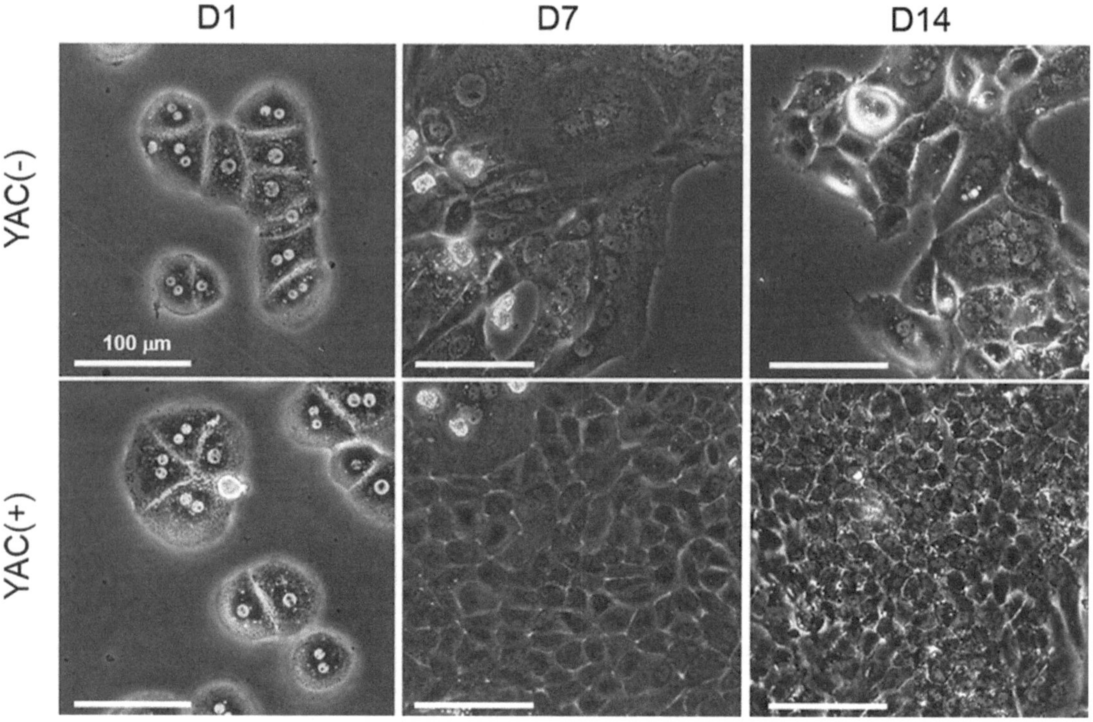

Fig. 1 Morphological change during reprogramming of MHs into CLiPs. MHs actively divide in the presence of YAC and become smaller in size and exhibit higher nucleus/cytoplasm ratio compared with MHs. This figure is reproduced from ref. [9]

3.3 Hepatic Induction of CLiPs

Hepatic differentiation is performed based on a previously reported protocol that induced hepatic maturation of fetal liver cells [13]. This protocol consists of two steps. In the first step, CLiPs are cultured for 6 days in SHM + YAC supplemented with 20 ng/mL of OSM and 10^{-6} M dexamethasone. In the second step, CLiPs are further cultured for another 2 days in the presence of Matrigel. Morphological changes in Step 1 are relatively mild. The nucleus/cytoplasm ratio of hepatic-induced cells (Hep-i(+) cells) is reduced, and the cytoplasm becomes rich in granules (Fig. 2a). However, morphological changes in Hep-i(+) cells become more evident in Step 2. Cell-cell boundaries become clear due to bile canaliculi formation (Fig. 2a). Hepatic differentiation can be confirmed by the expression of hepatic marker genes by qRT-PCR (Fig. 2b).

1. Harvest the cells following the steps described in Subheading 3.2, and obtain the required volume of the cell suspension for hepatic induction. We seed the cells at approximately 5×10^4 cells/cm^2 (1×10^5 cells/well and 2×10^5 cells/well for 24- and 12-well plates, respectively) (*see* **Note 11**). Centrifuge the tube at $200 \times g$ for 5 min.

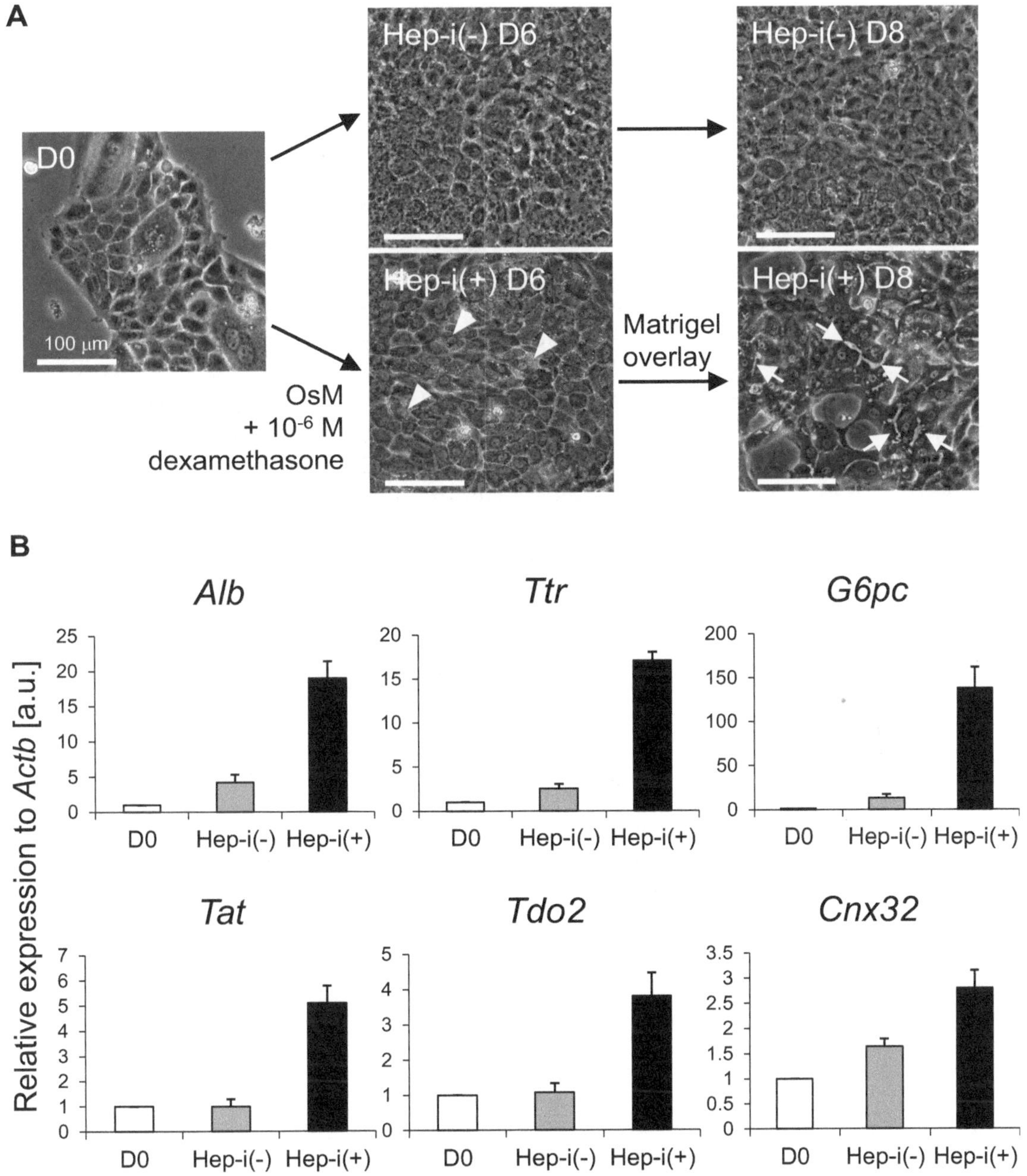

Fig. 2 Hepatic induction of CLiPs. (**a**) Morphological change of CLiPs during hepatic induction. Arrowheads indicate cells that are rich in granules in their cytoplasm. Arrows indicate bile canaliculi formed at the boundaries between cells that achieve hepatic maturation. (**b**) qRT-PCR of hepatic marker genes for Hep-i(+), Hep-i(−), and D0 cells. The data are shown as the mean ± S.E.M. of four independent experiments. This figure is partly reproduced from ref. [9]

2. Aspirate the supernatant, resuspend the cells in 0.5 or 1 mL SHM + YAC supplemented with 5% FBS, and seed to a well of collagen-coated 24- or 12-well plates, respectively (*see* **Note 12**).
3. One day after seeding, replace the medium with fresh SHM + YAC, and culture the cells for 2 days.
4. Replace the medium with HIM-1, and continue culturing for another 6 days. Replace the medium with fresh HIM-1 every other day.
5. Replace the medium with HIM-2, and continue culturing for another 2 days (*see* **Notes 13** and **14**).

3.4 Biliary Induction of CLiPs

CLiPs can be differentiated also into BECs using a two-step protocol (Fig. 3a). In Step 1, CLiPs are co-cultured on cell cycle-arrested mouse embryonic fibroblasts (MEFs) in mTeSR1 + YAC for 6 days. In Step 2, CLiPs are cultured for another 6 days in 2% Matrigel-containing mTeSR1 + YAC. CLiP-derived BEC-like cells form ductal and cystic structures typically observed in Step 2 (Fig. 3b). CLiP-derived ductal/cystic structures exhibit the capacity to transport water into their luminal space in response to secretin stimulation (Fig. 3c). Following incubation with fluorescein diacetate, these cells also transport fluorescein to their luminal spaces (Fig. 3d).

1. One day before harvesting CLiPs, seed cell cycle-arrested MEFs on 12-well collagen plates at 5×10^4 cells/well.
2. Harvest the primary CLiPs following the steps described in Subheading 3.2, and obtain the desired volume of the cell suspension for biliary induction. We seed the cells at 5×10^5 cells/well in 12-well plates. Centrifuge the tube at $200 \times g$ for 5 min.
3. Aspirate the supernatant, resuspend the cells in 1 mL SHM + YAC supplemented with 5% FBS, and seed onto the pre-inoculated MEFs (*see* **Note 12**).
4. (Step 1) On the following day, start biliary induction by replacing the medium with BIM-1. Culture the cells for 6 days while replacing the medium with fresh BIM-1 every other day.
5. (Step 2) Replace the medium with BIM-2. Culture the cells for another 6 days while replacing the medium with fresh BIM-2 every other day (*see* **Notes 14** and **15**).

3.5 Long-Term Culture of CLiPs Through Continuous Passages

CLiPs can be cultured stably through continuous passages; however, the proliferation rates transiently decrease from approximately passage 2 (P2) to 7 (P7) (*see* **Notes 16** and **17**). During long-term culture, the cells can be frozen for backup or future use (*see* Subheading 3.6). Primary CLiPs are typically passaged on days 14–16.

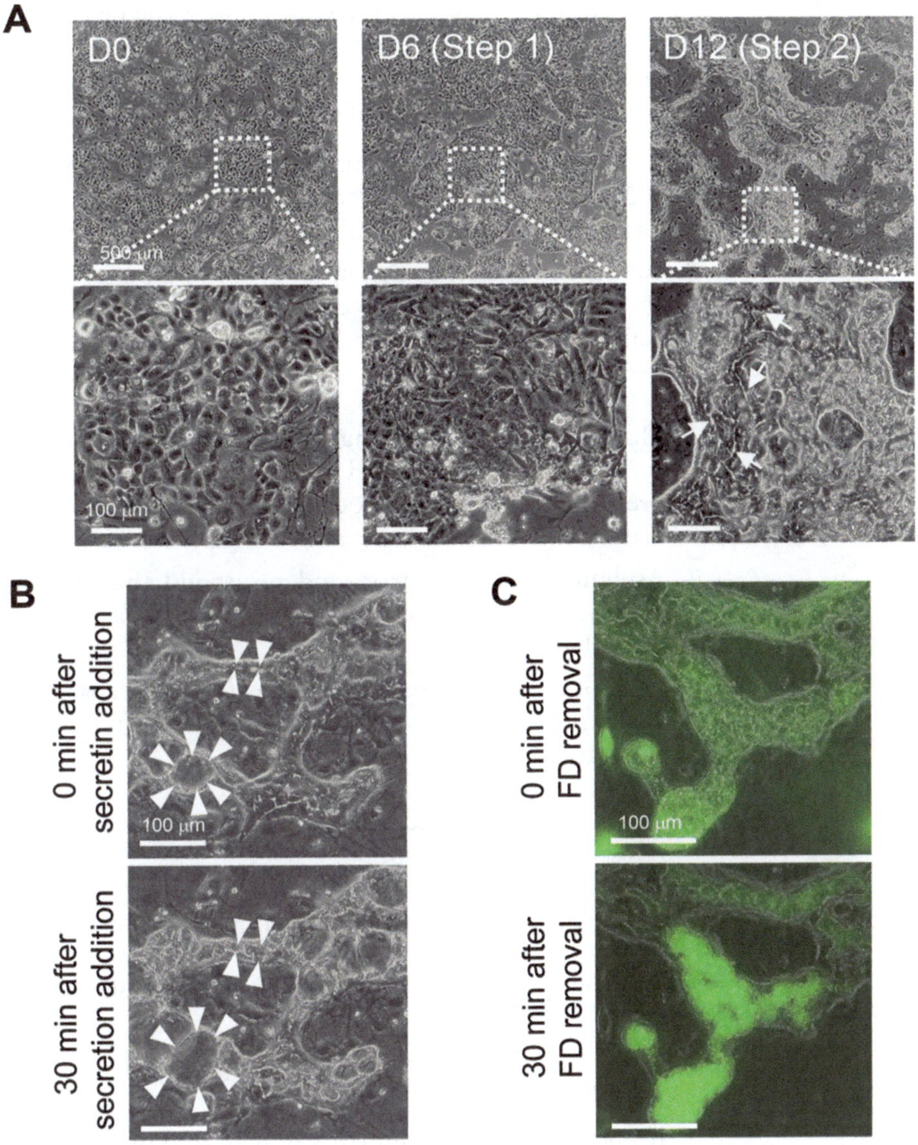

Fig. 3 Biliary induction of CLiPs. (**a**) Morphological change of CLiPs during biliary induction. Arrows indicate ductular structures formed between adjacent cells. (**b**) Phase-contrast images of BEC-i(+) cells taken before and 30 min after the addition of secretin. Arrowheads denote enlarged lumens following secretin treatment. (**c**) Uptake and excretion of fluorescein diacetate (FD) by the ductular structures of BEC-i(+) cells. Phase-contrast and fluorescence images were taken immediately after a 15-min incubation in the presence of FD (upper panel) and after another 30-min incubation in the absence of FD. This figure is partly reproduced from ref. [9]

Subcultured CLiPs are further passaged when they reach 50–90% confluence between P1 and P2, 30–60% confluence between P3 and P7, and 50–90% confluence thereafter. We typically seed cells at 2×10^6 viable cells/10 cm dish in earlier passages (P1 to P7). However, the actual seeded cell number is much lower than 2×10^6 viable cells/10 cm dish, because the cells in earlier passage cannot attach efficiently to the plates. Once cell proliferation rates increase (typically after approximately P5–7), we reduce the seeding density to 1–10 × 10^5 cells/10 cm dish so that they become subconfluent after 3–7 days of culture.

1. Harvest the cells following the steps described in Subheading 3.2, and obtain the desired volume of the cell suspension for passage. Centrifuge the tube at 200 × *g* for 5 min.
2. Resuspend the cells in an appropriate volume of SHM + YAC, seed on a desired plate coated with Matrigel, and culture the cells in a CO_2 incubator. Medium should be replaced every 1–3 days.

3.6 Freeze Stocking of CLiPs

1. Harvest the cells following the steps described in Subheading 3.2, and obtain a desired volume of the cell suspension for preparation of frozen stocks (typically 0.5–10 × 10^6 cells/vial). Centrifuge the tube at 200 × *g* for 5 min.
2. Aspirate the supernatant, and resuspend the cells in desired volume of Cell Banker. The volume of Cell Banker used is 0.5 mL/vial.
3. Aliquot to serum tubes equipped with screw caps. Seal the caps tightly, and store at −80 °C for at least 1 day. Then, transfer them to liquefied nitrogen.

4 Notes

1. Our centrifugation machine (Kubota 2800) uses 57 × *g* with the rotation speed at 600 rpm; researchers should adjust their rotation speed such that they can obtain 50–60 × *g* centrifugal force.
2. From this step forward, perform all procedures aseptically in a clean bench or a safety cabinet.
3. If undigested large fragments remain, remove them with sterilized tweezers.
4. When pipetting the MH suspension, a wide-bore pipette should be used to avoid damaging MHs by extensive shear stress.
5. When the gauze is clogged with undigested tissue fragments, remove the clogging materials from the filter using a pipette.
6. From this step forward, all the procedures should be performed on ice, and all the reagents should be kept at 4 °C or on ice.
7. Due to the heavy weight of MHs, be careful to keep the cell suspension homogeneous. Loading a large amount of the cell suspension in a pipette and dispensing it to multiple plates/wells may result in inconsistent seeding density among plates/

wells. We typically apply well-mixed undiluted cell suspension (generally approximately 3–8 × 10^6 cells/mL) using a micropipette directly to plates that are already filled with culture medium. During the cell seeding step, ensure that the cell suspension is kept homogeneous by periodically inverting the tube.

8. Make sure that the dissociating reagent completely covers the culture areas. We typically apply 2 mL of dissociation reagent to a 100 mm dish and use a similar ratio for other culture scales.
9. CLiPs are relatively resistant to trypsinization, and it typically takes 10–15 min to detach them from plates.
10. Despite its relatively damaging effect on the cells, this procedure is indispensable. Unlike general cell lines, even if the cells exhibit round-shaped morphology following trypsinization, many of them still attach to the plate tightly and can rarely be harvested only by adding medium onto the plates.
11. The cell density can be reduced to 1–2.5 × 10^4 cells/cm^2 when stable CLiPs are used for hepatic induction.
12. To increase the plating efficiency of primary CLiPs on culture plates, we recommend to use 5% FBS-supplemented SHM + YAC.
13. Prepare HIM-2 immediately prior to use.
14. Use P1000 or wide-bore 200 μL tips chilled at 4 °C or on ice.
15. Prepare BIM-2 immediately prior to use.
16. CLiPs undergo a transient decrease in their proliferative capacity between P2 and P7 (if not they enter complete senescence). Thus, during this period, it takes 15–40 days until the cells obtain 30–60% confluence, depending on experimental batches.
17. Prepare a Matrigel-coated plate before starting passage (we typically use 35 mm, 60 mm, and 10 cm dishes for cell maintenance).

Acknowledgments

This work was supported in part by Grant-in-Aid for Young Scientists B (16 K16643) and the Research Program on Hepatitis (16fk0310505h0005) from the Japan Agency for Medical Research and Development (AMED).

References

1. Woo DH, Kim SK, Lim HJ et al (2012) Direct and indirect contribution of human embryonic stem cell-derived hepatocyte-like cells to liver repair in mice. Gastroenterology 142:602–611. https://doi.org/10.1053/j.gastro.2011.11.0301
2. Oertel M, Menthena A, Chen Y-Q et al (2008) Purification of fetal liver stem/progenitor cells containing all the repopulation potential for normal adult rat liver. Gastroenterology 134:823–832. https://doi.org/10.1053/j.gastro.2008.01.007
3. Liu H, Kim Y, Sharkis S et al (2011) In vivo liver regeneration potential of human induced pluripotent stem cells from diverse origins. Sci Transl Med 3:82ra39. https://doi.org/10.1126/scitranslmed.3002376
4. Zhu S, Rezvani M, Harbell J et al (2014) Mouse liver repopulation with hepatocytes generated from human fibroblasts. Nature 508:93–97. https://doi.org/10.1038/nature13020
5. Sekiya S, Suzuki A (2011) Direct conversion of mouse fibroblasts to hepatocyte-like cells by defined factors. Nature 475:390–393. https://doi.org/10.1038/nature10263
6. Huang P, He Z, Ji S et al (2011) Induction of functional hepatocyte-like cells from mouse fibroblasts by defined factors. Nature 475:386–389. https://doi.org/10.1038/nature10116
7. Huch M, Dorrell C, Boj SF et al (2013) In vitro expansion of single Lgr5+ liver stem cells induced by Wnt-driven regeneration. Nature 494:247–250. https://doi.org/10.1038/nature11826
8. Lu W-Y, Bird TG, Boulter L et al (2015) Hepatic progenitor cells of biliary origin with liver repopulation capacity. Nat Cell Biol 17:971–983. https://doi.org/10.1038/ncb3203
9. Katsuda T, Kawamata M, Hagiwara K et al (2017) Conversion of terminally committed hepatocytes to culturable bipotent progenitor cells with regenerative capacity. Cell Stem Cell 20:41–55. https://doi.org/10.1016/j.stem.2016.10.007
10. Seglen PO (1976) Preparation of isolated rat liver cells. Methods Cell Biol 13:29–83. https://doi.org/10.1016/S0091-679X(08)61797-5
11. Loyer P, Cariou S, Glaise D et al (1996) Growth factor dependence of progression through G1 and S phases of adult rat hepatocytes in vitro. Evidence of a mitogen restriction point in mid-late G1. J Biol Chem 271:11484–11492
12. Frémin C, Bessard A, Ezan F et al (2009) Multiple division cycles and long-term survival of hepatocytes are distinctly regulated by extracellular signal-regulated kinases ERK1 and ERK2. Hepatology 49:930–939. https://doi.org/10.1002/hep.22730
13. Kamiya A, Kojima N, Kinoshita T et al (2002) Maturation of fetal hepatocytes in vitro by extracellular matrices and oncostatin M: induction of tryptophan oxygenase. Hepatology 35:1351–1359. https://doi.org/10.1053/jhep.2002.33331

Chapter 12

Induction of Functional Hepatocytes from Human iPSCs

Taketomo Kido and Yuta Koui

Abstract

Human-induced pluripotent stem cells (iPSCs) could be a useful source for production of hepatocytes. Here, we develop protocols to generate iPSC-derived liver progenitor cells, liver sinusoidal endothelial cells (LSECs), and hepatic stellate cells (HSCs). We also establish long-term two-dimensional co-culture system to induce functional hepatocytes from iPSC-derived liver cells.

Key words Human-induced pluripotent stem cells, Liver, Liver progenitor cells, Hepatocytes, Liver sinusoidal endothelial cells, Hepatic stellate cells

1 Introduction

Human-induced pluripotent stem cells (iPSCs) have been generated from a variety of somatic cells and used as an alternative cell source for production of different types of cells [1]. Several protocols have been reported for generation of hepatocytes from iPSCs [2–4]. However, induction of hepatocytes from iPSCs requires time-consuming multiple processes; also iPSC-derived hepatocytes exhibit immature phenotypes.

The liver consists of hepatocytes (parenchymal cells) and non-parenchymal cells, such as liver sinusoidal endothelial cells (LSECs) and hepatic stellate cells (HSCs). As they establish intercellular communications to make functional liver tissue during development [5, 6], co-culture of iPSC-derived hepatocytes and non-parenchymal cells would be useful for production of functional hepatocytes in vitro.

To establish this co-culture system, we have developed protocols to isolate iPSC-derived liver progenitor cells (LPCs), LSEC progenitors, and HSC progenitors based on the expression of cell surface specific markers (Table 1) [7, 8]. iPSC-derived LPCs and LSEC progenitors could be expanded by several passages and cryopreserved. To induce functional hepatocytes from iPSC-derived LPCs, we have also developed a high-density co-culture system [8].

Naoki Tanimizu (ed.), *Hepatic Stem Cells: Methods and Protocols*, Methods in Molecular Biology, vol. 1905, https://doi.org/10.1007/978-1-4939-8961-4_12, © Springer Science+Business Media, LLC, part of Springer Nature 2019

Table 1
Cell surface specific markers on liver cells

iPSC-derived liver cells	Specific markers	Reference
Liver progenitor cells (LPCs)	Carboxypeptidase M (CPM)	[7]
Liver sinusoidal endothelial cell (LSEC) progenitors	CD34/CD31 (PECAM1)/FLK1 (KDR, VEGFR2)	[8]
Hepatic stellate cell (HSC) progenitors	Activate leukocyte cell adhesion molecule (ALCAM)	[8]

2 Materials

1. Cell culture dishes/plates: 6 cm culture dish, 6-well plate, 12-well plate, 24-well plate, 48-well plate, Ultra-Low Attachment 6-well plate (Corning).
2. Accumax Cell Dissociation Solution (Innovative Cell Technologies).
3. StemPro Accutase Cell Dissociation Reagent (Thermo Fisher Scientific).
4. FcR Blocking Reagent (Miltenyi Biotec).
5. Cellmatrix Type I-C (Nitta Gelatin).
6. autoMACS Running Buffer (Miltenyi Biotec).
7. Antibodies: Anti-Carboxypeptidase M (CPM) antibody (Medical & Biological Laboratories Co., Ltd.), Purified Mouse IgG1 isotype control (BioLegend), PE Goat anti-mouse IgG antibody (BioLegend), FITC conjugated anti-CD34 antibody (BioLegend), PE conjugated anti-CD31 antibody (BD Bioscience), APC conjugated anti-CD309 (FLK1) antibody (Miltenyi Biotec), FITC mouse IgG1 isotype control (BD Bioscience), PE mouse IgG1 isotype control (BioLegend), APC mouse IgG1 isotype control (BD Bioscience), Anti-FITC microbeads (Miltenyi Biotec), Biotin conjugated anti-CD166 (ALCAM) antibody (Miltenyi Biotec), REA Control (S)-Biotin (Miltenyi Biotec), Streptavidin-APC (BD Bioscience).
8. Matrigel solution: 2% Matrigel (Growth Factor Reduced Basement Membrane Matrix, Thermo Fisher Scientific) in RPMI-1640 medium.
9. RPMI/B27 medium: 1% B-27 Supplement (Thermo Fisher Scientific) in RPMI-1640 medium.
10. Activin A/CHIR 99021 medium: 100 ng/mL Recombinant Activin A, 2 μM CHIR 99021 in RPMI/B27 medium.

11. Activin A medium: 100 ng/mL Recombinant Activin A in RPMI/B27 medium.
12. BMP-4/bFGF medium: 20 ng/mL Recombinant Human BMP-4, 10 ng/mL Recombinant Human FGF-basic (154 a. a.) in RPMI/B27 medium.
13. HGF medium: 20 ng/mL Recombinant Human HGF in RPMI/B27 medium.
14. Trypsin/EDTA solution: 0.05% Trypsin and 0.5 mM EDTA solution in D-PBS (−).
15. Mouse embryonic fibroblast (MEF) culture medium: 10% fetal bovine serum (FBS), 1% Penicillin-Streptomycin-Glutamine in DMEM high glucose.
16. BSA/PBS: 0.3% BSA in D-PBS (−).
17. Liver progenitor cell (LPC) culture medium: 10% FBS (JRH Biosciences), 1% Penicillin-Streptomycin-Glutamine (Thermo Fisher Scientific), 1% Insulin Transferrin, Selenium-X Solution (Thermo Fisher Scientific), 1% N-2 Supplement (Thermo Fisher Scientific), 1% MEM Non-Essential Amino Acids Solution (Thermo Fisher Scientific), 1% L-glutamine (Thermo Fisher Scientific), 1 mM L-ascorbic acid (Sigma-Aldrich Corporation), 10 mM nicotinamide (Wako Pure Chemical Industries), 2×10^{-7} M dexamethasone (Wako Pure Chemical Industries), 20 ng/mL Recombinant Human HGF (PeproTech), 10 ng/mL Recombinant Human EGF (PeproTech), 5 μM Y-27632 (Wako Pure Chemical Industries), 5 μM A83-01 (Tocris) in D-MEM/Ham's F-12 (Wako Pure Chemical Industries).
18. EDTA solution: 0.5 mM EDTA in D-PBS (−).
19. StemPro medium: StemPro-34 SFM (1×) (Thermo Fisher Scientific), StemPro-34 Nutrient Supplement (40×) (Thermo Fisher Scientific).
20. Fibronectin solution: 20 μg/mL human plasma fibronectin in D-PBS (−).
21. EC culture medium: EBM-2 Basal Medium (Lonza), EGM-2 SingleQuots (Lonza), 50 ng/mL Recombinant Human VEGF 165 (PeproTech).
22. LSEC culture medium: EBM-2 Basal Medium (Lonza), EGM-2 SingleQuots (Lonza), 50 ng/mL Recombinant Human VEGF 165 (PeproTech), 1.5 μM A83-01 (Tocris).
23. HSC culture medium: MSCBM (Lonza), MSCGM SingleQuots (Lonza), 10 μM Y-27632 (Wako Pure Chemical Industries).
24. Collagen gel solution: 10% 10 × conc. Ham F12 medium (Nitta Gelatin, KP-7100), 80% Cellmatrix Type I-A (Nitta

Gelatin, KP-2100), 10% Reconstitution buffer (50 mM NaOH, 260 mM $NaHCO_3$, 200 mM HEPES) (Nitta Gelatin, KP-8000).

25. Equipments: Ambient O_2/5% CO_2 incubator, 4% O_2/5% CO_2 incubator, autoMACS Pro Separator (Miltenyi Biotec), MoFlo XDP cell sorter (Beckman Coulter).

3 Methods

3.1 iPSC-Derived LPCs

3.1.1 Induction of LPCs from Human iPSCs

1. Preparation of culture plate: Add 1 mL of Matrigel solution to coat each well of a 6-well plate, and incubate the plate for 1 h at 37 °C.
2. Preparation of iPSCs: Remove the iPSC culture medium from each 6 cm culture dish, and wash one time with D-PBS (−) (*see* **Note 1**).
3. Aspirate D-PBS (−), add 1 mL of StemPro Accutase Cell Dissociation Reagent.
4. Incubate cells at 37 °C for 5 min. Harvest cells into 15 mL tube, and add iPSC culture medium for wash.
5. Centrifuge cells for 3 min at 340 × *g*. Resuspend cells in iPSC culture medium with 10 μM Y-27632.
6. Induction of LPCs from iPSCs: Remove the Matrigel solution immediately prior to use. Plate the cells at a density of $1–2 \times 10^4$ cells/cm^2, and incubate cells overnight at 37 °C with ambient O_2/5% CO_2 (Day 0) (*see* **Note 2**).
7. Replace medium with iPSC culture medium without Y-27632, and incubate cells overnight at 37 °C with ambient O_2/5% CO_2 (Day 1) (Fig. 1a).
8. Replace medium with Activin A/CHIR 99021 medium, and incubate cells for 2 days at 37 °C with ambient O_2/5% CO_2 (Day 2–3). Change medium every day.
9. Replace medium with Activin A medium, and incubate cells for 3 days at 37 °C with ambient O_2/5% CO_2 (Day 4–6) (Fig. 1b). Change medium every day.
10. Replace medium with BMP-4/bFGF medium, and incubate cells for 5 days at 37 °C with 4% O_2/5% CO_2 (Day 7–11) (Fig. 1c). Change medium every day.
11. Replace medium with HGF medium, and incubate cells for 10 days at 37 °C with 4% O_2/5% CO_2 (Day 12–22) (Fig. 1d). Change medium every day.

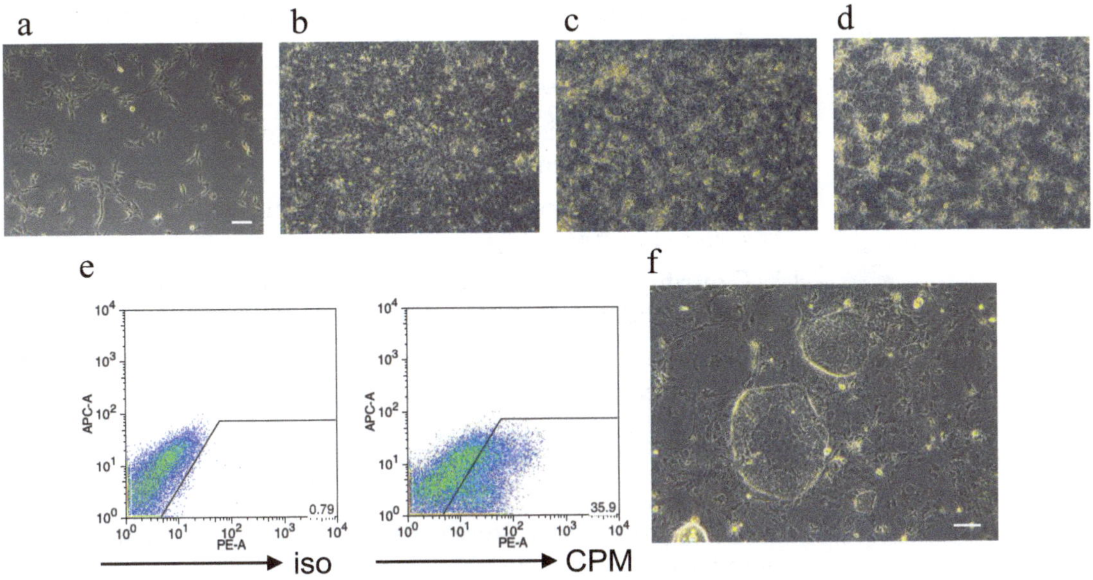

Fig. 1 Generation of LPCs from human iPSCs. (**a–d**) Morphological changes of iPSCs at different stages of hepatic differentiation. Scale bar, 100 μm. (**e**) FCM analysis of CPM expression. (**f**) Morphology of the CPM$^+$ LPC colonies on MEF feeder cells after 7 days of culture. Scale bar, 100 μm

3.1.2 Isolation and Expansion of iPSC-Derived LPCs

1. One day before isolation of iPSC-derived LPCs: Plate the mitomycin C-treated MEF cells at a density of 5–8 × 10^4 cells/cm^2 onto a 0.1% gelatin-coated 12-well plate in MEF culture medium. Incubate the cells overnight at 37 °C with ambient O_2/5% CO_2.
2. Remove the HGF medium from each well of 6-well plates, and wash once with D-PBS (−).
3. Aspirate D-PBS (−), add 0.5 mL of Trypsin/EDTA solution.
4. Incubate cells at 37 °C for 10–15 min. Harvest cells into 15 mL tube, and add MEF culture medium for wash.
5. Centrifuge cells for 3 min at 340 × *g*. Resuspend cells in MEF culture medium, and pass through 70 μm Cell Strainer to remove the debris.
6. Centrifuge cells for 3 min at 340 × *g*. Resuspend cells in 0.3% BSA/PBS.
7. Add FcR Blocking Reagent, and incubate the cells for 20 min on ice.
8. Add anti-Carboxypeptidase M (CPM) antibody, and incubate the cells for 30 min on ice. Use Purified Mouse IgG1 isotype control as negative control.
9. Add appropriate volume of 0.3% BSA/PBS for wash.
10. Centrifuge cells for 3 min at 340 × *g*. Resuspend cells in 0.3% BSA/PBS.

11. Add PE Goat anti-mouse IgG antibody, and incubate the cells for 30 min on ice (*see* **Note 3**).
12. Wash as in **step 9**.
13. Add propidium iodide to identify dead cells, and pass through 35 μm Cell Strainer.
14. Isolate iPSC-derived CPM^+ LPCs using a MoFlo XDP cell sorter (Fig. 1e).
15. Centrifuge iPSC-derived CPM^+ LPCs for 3 min at 340 × *g*. Resuspend cells in LPC culture medium.
16. Plate the cells at a density of 1–2 × 10^4 cells/cm^2 on mitomycin C-treated MEF feeder cells.
17. Incubate cells for 7–10 days at 37 °C with ambient O_2/5% CO_2. Change medium every day (Fig. 1f) (*see* **Note 4**).

3.2 iPSC-Derived LSECs

3.2.1 Induction of LSEC Progenitors from Human iPSCs

1. Remove the iPSC culture medium from each 6 cm culture dish, and wash once with D-PBS (−).
2. Aspirate D-PBS (−), add 1 mL of EDTA solution, and incubate cells at 37 °C for 5 min.
3. Aspirate EDTA solution, and add 3 mL of iPSC culture medium. Dissociate cells into small clusters, harvest cells into 15 mL tube, and add iPSC culture medium for wash (*see* **Note 5**).
4. Centrifuge cells for 3 min at 340 × *g*. Resuspend cells in StemPro medium with 10 μM Y-27632 and 2 ng/mL BMP-4.
5. Plate the cell clusters on Ultra-Low Attachment 6-well plates, and incubate cells overnight at 37 °C with 4% O_2/5% CO_2 (Day 0).
6. Harvest cells into 15 mL tube and centrifuge cells for 2 min at 120 × *g*. Resuspend cells in StemPro medium with 5 ng/mL Recombinant Activin A, 5 ng/mL Recombinant Human FGF-basic (154 a.a.), and 30 ng/mL BMP-4 (*see* **Note 6**).
7. Replate the cell clusters on Ultra-Low Attachment 6-well plates, and incubate cells for 3 days at 37 °C with 4% O_2/5% CO_2 (Day 1–4).
8. Harvest cells into 15 mL tube and centrifuge cells for 2 min at 120 × *g*. Resuspend cells in StemPro medium with 10 ng/mL Recombinant Human VEGF 165, 5.4 μM SB431542, and 0.5 μM Dorsomorphin dihydrochloride (*see* **Note 6**).
9. Replate the cell clusters on Ultra-Low Attachment 6-well plates, and incubate cells for 2 days at 37 °C with 4% O_2/5% CO_2 (Day 4–6).
10. Harvest cells into 15 mL tube and centrifuge cells for 2 min at 120 × *g*. Resuspend cells in EC culture medium.

11. Replate the cell clusters at a 1:2 split ratio onto a 0.1% gelatin-coated 6-well plate.
12. Incubate the cells for 7 days at 37 °C with 4% O_2/5% CO_2. Change medium every other day (*see* **Note 7**).

3.2.2 Isolation and Expansion of iPSC-Derived LSEC Progenitors

1. Remove the EC culture medium from each well of 6-well plates, and wash once with D-PBS (−).
2. Aspirate D-PBS (−), and add 0.5 mL of Trypsin/EDTA solution.
3. Incubate cells at 37 °C for 10 min. Harvest cells into 15 mL tube, and add MEF culture medium for wash.
4. Centrifuge cells for 3 min at 340 × *g*. Resuspend cells in 0.3% BSA/PBS, and pass through 70 μm Cell Strainer to remove the debris.
5. Centrifuge cells for 3 min at 340 × *g*. Resuspend cells in 0.3% BSA/PBS.
6. Add FcR Blocking Reagent, and incubate the cells for 20 min on ice.
7. Add FITC conjugated anti-CD34 antibody, PE conjugated anti-CD31 antibody, and APC conjugated anti-CD309 (FLK1) antibody, and incubate the cells for 30 min on ice. Use FITC mouse IgG1 isotype control, PE mouse IgG1 isotype control, and APC mouse IgG1 isotype control as negative control.
8. Add appropriate volume of 0.3% BSA/PBS for wash.
9. Centrifuge cells for 3 min at 340 × *g*. Resuspend cells in autoMACS Running Buffer.
10. Add Anti-FITC microbeads and incubate the cells for 20 min on ice.
11. Add appropriate volume of autoMACS Running Buffer for wash.
12. Centrifuge cells for 3 min at 340 × *g*. Resuspend cells in autoMACS Running Buffer.
13. Enrich $FITC^+$ cells using autoMACS Pro Separator (Miltenyi Biotec), and centrifuge cells for 3 min at 340 × *g*. Resuspend cells in 0.3% BSA/PBS.
14. Add propidium iodide to identify dead cells, and pass through 35 μm Cell Strainer.
15. Isolate iPSC-derived $CD34^+CD31^+FLK1^+$ LSEC progenitors using a MoFlo XDP cell sorter (Fig. 2).
16. Centrifuge iPSC-derived $CD34^+CD31^+FLK1^+$ LSEC progenitors for 8 min at 340 × *g*. Resuspend cells in EC culture medium.

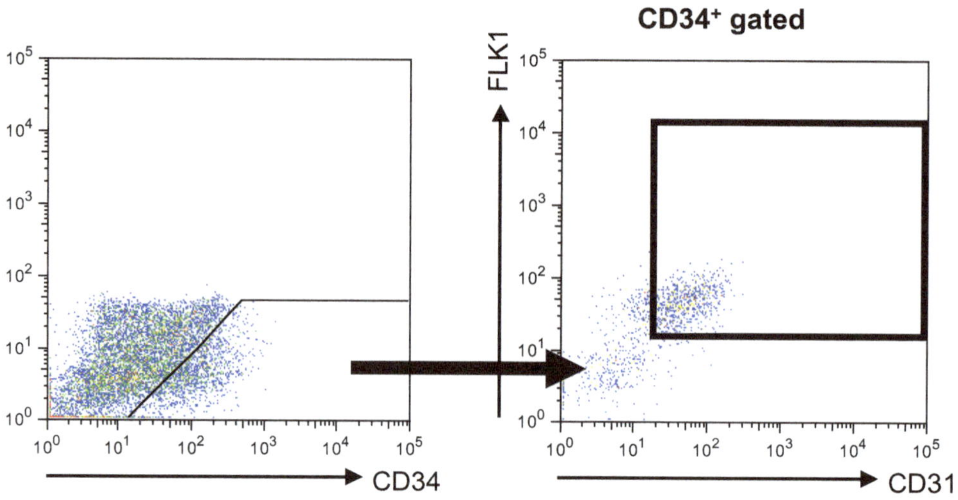

Fig. 2 Generation of LSEC progenitors from human iPSCs. FCM analysis of iPSC-derived LSEC progenitors after enrichment of CD34$^+$ cells using autoMACS Pro Separator. CD34$^+$ cells were identified (left), and CD31$^+$FLK1$^+$ cells were identified in CD34$^+$ cell fraction

17. Plate the cells at a density of 2×10^4 cells/cm^2 on 12-well plates coated with Fibronectin solution.
18. Incubate cells for 7 days at 37 °C with 4% O_2/5% CO_2. Change medium every other day (*see* **Note 8**).

3.2.3 Induction of iPSC-Derived Mature LSECs from LSEC Progenitors

1. Remove the EC culture medium from each well of 12-well plates, and wash once with D-PBS (−).
2. Aspirate D-PBS (−), and add 0.5 mL of Trypsin/EDTA solution.
3. Incubate cells at 37 °C for 10 min. Harvest cells into 15 mL tube, and add MEF culture medium for wash.
4. Centrifuge cells for 3 min at 340 × *g*. Resuspend cells in LSEC culture medium.
5. Plate the cells at a density of 1.5×10^4 cells/cm^2 on 12-well plates coated with Fibronectin solution.
6. Incubate cells for 7 days at 37 °C with 4% O_2/5% CO_2. Change medium every other day.
7. Passage as in Subheading 3.2.3, **steps 1–5**.
8. Incubate cells for 7 days at 37 °C with 4% O_2/5% CO_2. Change medium every other day (*see* **Note 9**).

3.3 iPSC-Derived HSCs

3.3.1 Induction of HSC Progenitors from Human iPSCs

Transfer the iPSCs to HSC progenitor differentiation culture as in Subheading 3.2.1, **steps 1–9**.

3.3.2 Isolation of HSC Progenitors and Induction of iPSC-Derived Mature HSCs

1. Harvest cells into 15 mL tube and centrifuge cells for 3 min at 340 × *g*.
2. Aspirate the medium and resuspend cells in Accumax Cell Dissociation Solution.
3. Incubate cells at 37 °C water bath for 10 min.
4. Add MEF culture medium to inactivate dissociation solution.
5. Centrifuge cells for 3 min at 340 × *g*. Resuspend cells in 0.3% BSA/PBS, and pass through 70 μm Cell Strainer to remove the debris.
6. Centrifuge cells for 3 min at 340 × *g*. Resuspend cells in 0.3% BSA/PBS.
7. Add FcR Blocking Reagent, and incubate the cells for 20 min on ice.
8. Add Biotin conjugated anti-CD166 (ALCAM) antibody, and incubate the cells for 30 min on ice. Use REA Control (S)-Biotin as negative control.
9. Add appropriate volume of 0.3% BSA/PBS for wash.
10. Centrifuge cells for 3 min at 340 × *g*. Resuspend cells in 0.3% BSA/PBS.
11. Add Streptavidin-APC, and incubate the cells for 20 min on ice.
12. Wash as in **step 9**.
13. Add propidium iodide to identify dead cells, and pass through 35 μm Cell Strainer.
14. Isolate iPSC-derived $ALCAM^{high}$ HSC progenitors using a MoFlo XDP cell sorter (Fig. 3).
15. Centrifuge iPSC-derived $ALCAM^{high}$ HSC progenitors for 3 min at 340 × *g*. Resuspend cells in HSC culture medium.
16. Plate the cells at a density of 1.5×10^4 cells/cm^2 on Cellmatrix Type I-C-coated 24-well plates.
17. Incubate cells for 5 days at 37 °C with ambient O_2/5% CO_2. Change medium every other day (*see* **Note 10**).

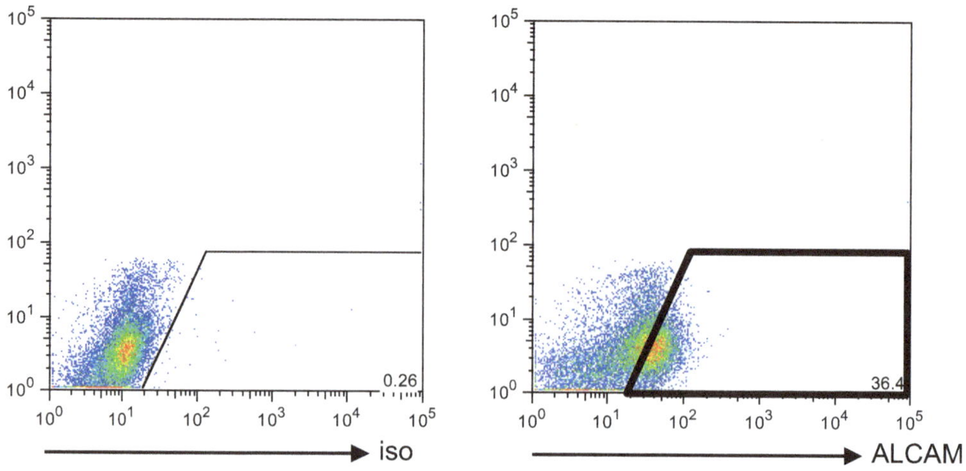

Fig. 3 Generation of HSC progenitors from human iPSCs. FCM analysis of ALCAM expression in iPSC-derived mesodermal cells (right). The positive gate was defined by the isotype control (left)

3.4 High-Density Co-culture System by iPSC-Derived LPCs, LSECs, and HSCs

3.4.1 Preparation of iPSC-Derived LPCs

1. Remove the medium from LPC culture plates (from Subheading 3.1.2, **step 17**), and wash once with D-PBS (−).
2. Aspirate D-PBS (−), add 0.5 mL of Trypsin/EDTA solution.
3. Incubate cells at 37 °C for 10–15 min. Harvest cells into 15 mL tube, and add MEF culture medium for wash.
4. Centrifuge cells for 3 min at 340 × *g*. Resuspend cells in MEF culture medium, and pass through 70 μm Cell Strainer to remove the debris.
5. Centrifuge cells for 3 min at 340 × *g*. Resuspend cells in MEF culture medium.
6. Plate the cells onto a 0.1% gelatin-coated 10 cm culture dish. Incubate cells for 30 min to remove adherent MEF cells (*see* **Note 11**).
7. Harvest non-adherent cells (iPSC-derived LPCs) into 15 mL tube and centrifuge for 3 min at 340 × *g*. Resuspend cells in LPC culture medium.

3.4.2 Preparation of iPSC-Derived LSECs and HSCs

1. Remove the medium from each culture plate (Subheading 3.2.3, **step 8**, and Subheading 3.3.2, **step 17**), and wash once with D-PBS (−).
2. Aspirate D-PBS (−), and add 0.5 mL of Trypsin/EDTA solution.
3. Incubate cells at 37 °C for 5–10 min. Harvest cells into 15 mL tube, and add MEF culture medium for wash.
4. Centrifuge cells for 3 min at 340 × *g*. Resuspend cells in LPC culture medium.

3.4.3 Co-culture of iPSC-Derived LPCs, LSECs, and HSCs

1. Preparation of collagen gel-coated plates: Add 0.2 mL of collagen gel solution to coat each well of a 48-well plate, and incubate the plate for 30 min at 37 °C.
2. Plate iPSC-derived LPCs (2×10^5 cells/well), iPSC-derived LSECs (2×10^4 cells/well), and iPSC-derived HSCs (2×10^4 cells/well) on collagen gel-coated 48-well plates.
3. Incubate cells for 10 weeks at 37 °C with ambient O_2/5% CO_2. Change medium every other day (*see* **Note 12**).

4 Notes

1. When iPSCs reached subconfluence, they can be used for induction of LPCs.
2. The cell density depends on the growth rate and varies between the iPSC lines.
3. It is possible to choose different fluorescent-dye conjugated secondary antibodies.
4. The expanded cells could be cryopreserved for further study.
5. Pipette slowly to prevent excessive dissociation of iPSC clusters. Excessive dissociation results in decrease of differentiation efficiency.
6. Pipette slowly because the cell clusters break easily.
7. Cell clusters would adhere on the surface of the plate. Then, the outgrowth of cells would be observed surrounding the adherent cell clusters.
8. LSEC progenitors would reach confluence and exhibit endothelial cell morphology after 7 days of culture.
9. A83-01 (TGFβRI inhibitor) is a key molecule for inducing LSEC maturation. After culturing in the presence of A83-01, the expression levels of the mature LSEC markers, FCGR2B, STAB2, F8, and LYVE1, would be highly upregulated.
10. Y27632 (ROCK inhibitor) is a key factor for supporting HSC maturation. iPSC-derived HSCs proliferate and exhibit typical mature HSC morphology with projections after 5 days of culture in the presence of Y27632. We can observe vitamin A droplets in iPSC-derived HSCs after incubation with vitamin A.
11. After 30 min of incubation, we can observe adherent MEF cells under the microscope.
12. Measure CYP3A4 activity every week. It would gradually increase during 10 weeks of culture (Fig. 4).

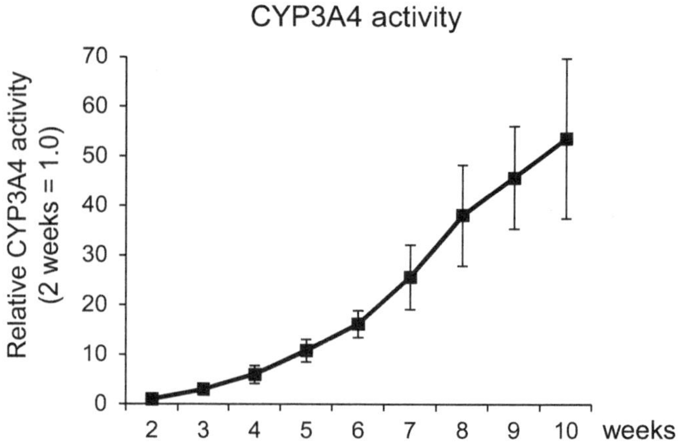

Fig. 4 CYP3A4 activity. Relative Cytochrome P450 3A4 (CYP3A4) activity in iPSC-derived liver cells. The results are shown as the mean ± SEM of three independent experiments. The measurements were performed every week

Acknowledgment

This study was supported by CREST program of Japan Science and Technology Agency, Grants-in-Aid for Scientific Research of Japan Society for the Promotion of Science, and Japan Agency for Medical Research and Development (AMED).

References

1. Takahashi K, Tanabe K, Ohnuki M, Narita M, Ichisaka T, Tomoda K, Yamanaka S (2007) Induction of pluripotent stem cells from adult human fibroblasts by defined factors. Cell 131:861–872
2. Si-Tayeb K, Noto FK, Nagaoka M, Li J, Battle MA, Duris C, North PE, Dalton S, Duncan SA (2010) Highly efficient generation of human hepatocyte-like cells from induced pluripotent stem cells. Hepatology 51:297–305
3. Takayama K, Inamura M, Kawabata K, Sugawara M, Kikuchi K, Higuchi M, Nagamoto Y, Watanabe H, Tashiro K, Sakurai F, Hayakawa T, Furue MK, Mizguchi H (2012) Generation of metabolically functioning hepatocytes from human pluripotent stem cells by FOXA2 and HNF1a transduction. J Hepatol 57:628–636
4. Ogawa S, Surapisitchat J, Virtanen C, Ogawa M, Niapour M, Sugamori KS, Wang S, Tamblyn L, Guillemette C, Hoffmann E, Zhao B, Strom S, Laposa RR, Tyndale RF, Grant DM, Keller G (2013) Three-dimensional culture and cAmp signaling promote the maturation of human pluripotent stem cell-derived hepatocytes. Development 140:3285–3296
5. Matsumoto K, Yoshitomi H, Rossant J, Zaret KS (2001) Liver organogenesis promoted by endothelial cells prior to vascular function. Science 294:559–563
6. Yin C, Evason KJ, Asahina K, Stainier DY (2013) Hepatic stellate cells in liver development, regeneration, and cancer. J Clin Invest 123:1902–1910
7. Kido T, Koui Y, Suzuki K, Kobayashi A, Miura Y, Chern EY, Tanaka M, Miyajima A (2015) CPM is a useful cell surface marker to isolate expandable bi-potential liver progenitor cells derived from human iPS cells. Stem Cell Rep 5:508–515
8. Koui Y, Kido T, Ito T, Oyama H, Chen SW, Katou Y, Shirahige K, Miyajima A (2017) An in vitro human liver model by iPSC-derived parenchymal and non-parenchymal cells. Stem Cell Rep 9:490–498

Chapter 13

Culture System of Bile Duct-Like Cystic Structures Derived from Human-Inducible Pluripotent Stem Cells

Akihide Kamiya, Kazuya Anzai, Kota Tsuruya, and Hiromi Chikada

Abstract

Inducible pluripotent stem (iPS) cells are multipotent stem cells that are produced by gene transfer of reprogramming factors to somatic cells. They are thought to be an important source of regenerative medicine because of their pluripotency and self-renewal ability. Although the liver has high regeneration ability, continuous death of hepatocytes due to chronic inflammation leads to liver cirrhosis and liver carcinoma. With regard to such serious liver diseases, liver transplantation is used as a complete cure, but there is a problem of donor shortage. Therefore, transplantation therapy using liver tissue generated from stem cells in vitro is expected.

We are developing a system to induce the differentiation of cholangiocytes, one of important non-parenchymal cells in living liver tissue, from human iPS cells. Bile duct-like cystic structures can be induced by purifying human iPS cell-derived hepatoblasts expressing hepatic progenitor cell surface markers and inducing differentiation under appropriate culture conditions. These cells are considered to be useful in constructing a hepatic organoid that reproduces the liver structure of the living body.

Key words Hepatic progenitor cells, Human iPS cells, Induction of hepatic differentiation, Long-term proliferation, Purification of progenitor cell

1 Introduction

The liver is the largest organ in the body and contains various functions for maintaining homeostasis in the body. Hepatocytes (liver parenchymal cells) express various genes for mature liver function such as synthesis of serum proteins, metabolism of carbohydrates and lipids, and detoxification of alcohol and drugs. In addition, several types of non-parenchymal cells are also present in the liver. Cholangiocytes constitute the intrahepatic bile ducts. Stellate cells are involved in the accumulation of vitamin A. Kupffer cells are intrahepatic macrophages, and sinusoidal endothelial cells constitute liver sinusoidal vessels in the liver. It is known that cell-cell interactions between these non-parenchymal cells and hepatocytes are important for liver function. Cholangiocytes

Naoki Tanimizu (ed.), *Hepatic Stem Cells: Methods and Protocols*, Methods in Molecular Biology, vol. 1905, https://doi.org/10.1007/978-1-4939-8961-4_13,

constitute the bile duct system for excretion of bile acid derived from hepatocytes [1]. Bile acids synthesized by hepatocytes are released from the apical side of the hepatocyte membrane to the bile canaliculus and are excreted into the small intestine and gall-bladder through the intrahepatic bile duct network formed by cholangiocyte. Therefore, disorders of the intrahepatic bile ducts such as primary biliary cholangitis lead to poor elimination of bile acid from the liver and cause various liver injuries.

During embryonic development, the ventral foregut endoderm begins to form early liver buds by migrating into the septum transversum (embryonic day 8–9 in mice) [2, 3]. Subsequently, hepatic buds grow by soluble factor signals from surrounding tissues, and hepatic progenitor cells (hepatoblasts), which are the cells that differentiate into the mature hepatocytes and bile duct cells, grow inside. Hepatoblasts surrounding the portal vein form ductal plate in the middle of embryogenesis (13–15 days of mouse embryo) [4]. Part of ductal plate cells lose hepatocytic gene expression and form a luminal structure of bile duct. Several studies using model mice describe that the cell-cell interaction between hepatoblasts and mesenchymal cells around the portal vein and the signal of soluble factors are important for differentiation from hepatoblasts to cholangiocytes. For example, the concentration of transforming growth factor β (TGFβ) in the liver during mid-fetal embryos is high around the portal vein and low as it approaches the inside of the liver parenchyma. The concentration gradient of TGFβ induces differentiation of hepatoblasts [5]. In addition, Notch-Jagged signal is important for bile duct formation process, and Jagged-1 is the causative gene of Alagille syndrome accompanied by abnormalities of the bile duct system [6, 7]. It is also known that various transcription regulators are important for bile duct formation and differentiation. Sal-like protein 4 is a transcription factor that regulates the fate determination of hepatoblasts and promotes bile duct differentiation by inducing cholangiocytic gene expression [8]. It has also been reported that grainyhead-like 2 regulates bile duct formation via gene expression control of miR122 [9].

Complex three-dimensional structure of the liver is crucial for its functions. In order to reproduce the liver functional structure in vitro, organoid culture method has been applied to co-culture of hepatocytes, mesenchymal cells, and endothelial cells [10, 11]. However, these hepatic organoids do not contain the bile duct structure, and the cytotoxicity of bile acids synthesized by hepatocytes against cells in these organoids is a potential problem. Therefore, the construction of a novel hepatic organoid using a co-culture system containing cholangiocytes is required. Various groups described about the in vitro culture for induction of cholangiocytes. We previously established an in vitro amplification system of hepatic progenitor cells derived from human-inducible pluripotent stem (iPS) cells by purifying $CD13^+CD133^+$ cells after sequential treatment of cytokines. In addition, bile duct-like cystic structures can be induced by embedding these hepatic progenitor cells in extracellular matrix gel and performing three-dimensional culture [12]. Other groups also reported differentiation system of human iPS cells into bile ductal cells by changing culture conditions [13, 14].

Polycystic liver diseases (PLD), which contain multiple cysts in the liver, are caused by genetic abnormalities of cholangiocytes. Recent analyses revealed that several genes such as protein kinase C substrate 80 K-H and SEC63 are involved in the onset of PLD [15, 16]. Cilium formation is regulated by these genes, but the molecular mechanism causing the pathological condition of PLD is still unknown [17]. Advances in genome editing technology have made it possible to introduce gene mutations in human iPS cells. Therefore, it is possible to induce mutation in the causative gene of bile duct disease in human iPS cells, to induce differentiation into hepatic progenitor cells and cholangiocytic cells. These cells are useful to reproduce the pathology of PLD or other bile duct-related diseases in vitro and screen therapeutic drugs. In this chapter, we introduce our culture system inducing bile duct-like tissues derived from human iPS cells.

2 Materials

2.1 Differentiation of Human iPS Cells into Hepatic Progenitor Cells

The approval of a suitable committee is usually required for animal experiments.

1. Human iPS cells: HiPS-RIKEN-2F human iPS cells are established from human umbilical cord-derived fibroblasts by introduction of Oct3/4, Klf4, Sox2, and c-Myc using a retroviral system [18]. These cells are cryopreserved in liquid nitrogen using DAP213 cryopreserve solution (*see* **Note 1**).
2. 0.1% gelatin: Gelatin derived from porcine skin is dissociated with PBS and autoclaved.
3. Fetal bovine serum (FBS): FBS is inactivated by incubation at 55 °C for 30 min.
4. Mouse embryonic fibroblast (MEF) culture medium: Dulbecco's modified Eagle's medium (DMEM) supplemented with 10% FBS and 1× penicillin–streptomycin–glutamine.
5. Phosphate-buffered saline (PBS): 10× PBS without $CaCl_2$ and $MgCl_2$ is diluted with ultrapure water, autoclaved, and stored at 4 °C.
6. Mitomycin C: A 200× stock solution (2 mg/mL) is prepared with ultrapure water.

7. Human iPS cell medium: 0.1 mM nonessential amino acids, 1× penicillin–streptomycin–glutamine, 20% knockout serum replacement, 0.1 mM 2-mercaptoethanol, 5 ng/mL recombinant human basic fibroblast growth factor (FGF) in DMEM/F12 medium.
8. Differentiation medium: RPMI 1640 containing 2% B27 supplement.
9. Activin A (1000× stock solution): 100 mg/mL recombinant human activin A in PBS with 0.1% bovine serum protein (BSA).
10. Basic FGF (1000× stock solution): 10 mg/mL recombinant human basic FGF in PBS with 0.1% BSA.
11. Bone morphogenetic protein (BMP)-4 (1000× stock solution): 20 mg/mL recombinant human BMP-4 in PBS with 0.1% BSA.
12. Recombinant human hepatocyte growth factor (HGF) (1000× stock solution): 40 mg/mL HGF in PBS with 0.1% BSA.
13. Staining medium: PBS with 0.3% FBS.
14. Stock propidium iodide solution (1000×): 1 mg/mL propidium iodide in PBS. The stock solution is diluted with staining medium.
15. Antibodies for flow cytometry: *see* Table 1.

Table 1
List of antibodies used for flow cytometry experiments and immunocytochemistry for human iPS cell differentiation and cholangiocyte differentiation

Primary antibodies for flow cytometry	Clone	Source	Catalog number
CD13-PE	WM15	BD Pharmingen	555394
CD133/1-APC	AC133	Miltenyi Biotec	130–090-826
Primary antibodies for immunostaining	**Dilution**	**Source**	**Catalog number**
α-Fetoprotein (AFP) (C3)	1/600	Dako	A0008
Keratin-7 (CK7)	1/1000	Dako	M7018
β-Catenin	1/1000	BD Pharmingen	610154
F-Actin (Acti-stain 488 phalloidin)	1/150	Cytoskeleton, Inc.	PHDG1
Protein kinase Cζ (C-20)	1/500	Santa Cruz	sc-216
Integrin α6 (CD49f) (GoH3)	1/500	BD Pharmingen	555734
Secondary antibodies	**Dilution**	**Source**	**Catalog number**
Anti-mouse/Alexa Fluor 555	1/1000	Invitrogen	A31570
Anti-rat/Alexa Fluor 488	1/1000	Invitrogen	A21208
Anti-rabbit/Alexa Fluor 488	1/1000	Invitrogen	A21206

2.2 Expansion of Hepatic Progenitor Cells and Induction of Cholangiocytic Differentiation Using Extracellular Matrices

1. H-CFU-C medium: 1× Insulin–Transferrin–Selenium X, 10 mM nicotinamide, 2.5 mM HEPES buffer solution, 1× penicillin–streptomycin–glutamine, 0.1 mM nonessential amino acids.
2. Standard culture medium: 1:1 mixture of hepatic colony-forming unit (H-CFU-C) medium and DMEM with 10% FBS, 10^{-7} M dexamethasone, 0.25 μM A-83-01, 10 μM Y-27632, 40 ng/mL recombinant human HGF, and 20 ng/mL recombinant human epidermal growth factor (EGF).
3. Extracellular matrix gel solution: A mixture of 40% collagen type-IA solution (Nitta gelatin) and 40% matrigel in DMEM/F12 medium.
4. Cholangiocyte culture medium: 2% B27 supplement, 0.25 μM A-83-01, 10 μM Y-27632, 20 ng/mL EGF, 40 ng/mL HGF, 40 ng/mL recombinant human Wnt-3a, and 100 ng/mL recombinant human R-spondin 1 in 1:1 mixture of H-CFU-C medium and DMEM/F-12.

2.3 Characterization of Human iPS Cells-Derived Cholangiocytic Cells

2.3.1 Analyses of Cholangiocytic Differentiation Using RT-PCR

1. RNAiso Plus: Solution for RNA purification.
2. ReverTra Ace® qPCR RT Master Mix: Solution for cDNA synthesis.
3. THUNDERBIRD® Probe qPCR Mix: Solution for quantitative PCR.
4. Universal Library: Detection probes for quantitative PCR. The probe numbers and primer sequences for each gene are provided in Table 2.

2.3.2 Analyses of Cholangiocytic Differentiation Using Immunocytochemistry

1. PBS+: PBS containing 1 mM $MgCl_2$ and 1 mM $CaCl_2$.
2. Collagenase-1: 1000 U/mL high-purity collagenase in PBS+. Aliquots are stored at −80 °C Thaw and dilute 1:10 in PBS+ before use.
3. 4% paraformaldehyde: Solution for cell fixation.
4. Blocking medium: 5% donkey serum and 0.5% Triton-X100 solution in PBS+. Donkey serum is used after heat inactivation by incubation at 56 °C for 30 min.
5. 4′,6-Diamidino-2-phenylindole dihydrochloride (DAPI) solution: A 1000× solution (2 mg/mL) is prepared with methanol.
6. Immunofluorescent mount solution: Dako Fluorescence Mounting Medium.

Table 2
PCR primers for detection of human gene expression

Human genes	Forward primer (5′–3′)	Reverse primer (5′–3′)	Probe number
AFP	tgtactgcagagataagtttagctgac	tccttgtaagtggcttcttgaac	61
GRHL2	ggacagcacatacagcgaga	agccccaactgaagcactc	5
HNF4α	attgacaacctgttgcagga	cgttggttcccatatgttcc	3
HPRT1	tgaccttgatttattttgcatacc	cgagcaagacgttcagtcct	73
Keratin 7	ctgaggctgaagcctggta	gggtattccggaggtcgt	64
Keratin 19	gccactactacacgaccatcc	caaacttggttcggaagtcat	71

Afp α-feto protein, *Grhl* grainyhead like transcription factor, *HNF* hepatocyte nuclear factor, *HPRT1* hypoxanthine phosphoribosyltransferase 1

3 Methods

3.1 Differentiation of Human iPS Cells into Hepatic Progenitor Cells

Methods for maintaining human iPS cells and differentiation into hepatic progenitor cells were already shown in "Generation and in vitro expansion of hepatic progenitor cells from human iPS cells" in detail [19].

1. Embryonic day (E) 12 ICR mouse embryos were dissected, and the head and internal organs are surgically removed. The torso is minced and dissociated in 0.05% trypsin–EDTA. After centrifugation and washing steps, cells were inoculated on culture dishes with MEF culture medium. MEFs are incubated with 0.01 mg/mL mitomycin C in MEF culture medium at 37 °C for 2 h.
2. Human iPS cells are cultured and maintained on mitomycin C-treated MEFs. The cells are cultured at 37 °C in a 5% CO_2 incubator. Human iPS cells are proliferated into Semi-confluent state and used for hepatocytic differentiation.
3. Semi-confluent human iPS cells are cultured in Differentiation medium supplemented with 100 ng/mL activin A on days 0–4 at 37 °C in a 5% CO_2 incubator.
4. After activin A stimulation, iPS cells were cultured in Differentiation medium supplemented with 10 ng/mL basic FGF and 20 ng/mL BMP-4 on days 4–8.
5. After FGF and BMP-4 stimulation, iPS cells were cultured in Differentiation medium supplemented with HGF on days 8–12. These cells are used for the isolation of hepatic progenitor cells.
6. After 12 days of culture, cells are washed with PBS and dissociated with 0.05% trypsin–EDTA (1 mL/60 mm dish). Trypsinized cells are collected into a 15 mL tube and centrifuged

(260 × *g* for 5 min). Fluorescent-conjugated antibodies are added to the cell pellets (Antibody aliquots: APC-conjugated anti-CD133 and PE-conjugated anti-CD13, Table 1) and incubated at 4 °C for 1 h (*see* **Note 2**).

7. After the washing step, cells were stained with propidium iodide and analyzed using flow cytometer.
8. Doublets are excluded by FSC, SSC, and pulse width. Dead cells are excluded by propidium iodide staining. $CD13^{high}CD133^{+}$ cells are sorted as a hepatic progenitor cell fraction (*see* **Note 3**).

3.2 Culture of Hepatic Progenitor Cells and Induction of Cholangiocytic Differentiation Using Extracellular Matrices

1. Mitomycin C-treated MEF cells (2×10^5 cells/well) are plated onto 0.1% gelatin-coated 12-well plates and maintained in MEF culture media the day before sorting. These cells are incubated for at least 12 h at 37 °C in a CO_2 incubator.
2. Sorted $CD13^{high}CD133^{+}$ cells (derived from Subheading 3.1) are cultured on MEF cells. Standard culture medium is used for cell culture (750–1000 μL per well). Medium is replaced every 3 days.
3. Individual $CD13^{high}CD133^{+}$ cell clonally proliferates and forms a hepatic progenitor colony on MEF for 10–14 days of culture. The colonies are dissociated by 5 min incubation of 0.05% trypsin–EDTA at 37 °C. After incubation, 1 mL of DMEM containing 10% FBS is added, and cells are collected by the centrifugation at 260 × *g* for 5 min.
4. Cells are replated onto new mitomycin C-treated MEF cells. Approximately 1/10 to 1/3 of the cells are seeded on individual well. After 7–10 days of culture, the passaged cells form new colonies.
5. Several passaged colonies derived from $CD13^{high}CD133^{+}$ cells are dissociated using 0.05% trypsin–EDTA and collected in DMEM containing 10% FBS as shown above. The cell number is counted.
6. The cells are then combined with an extracellular matrix gel solution (1500 cells/50 μL extracellular matrix gel/well). After mixing by pipetting, cells embedded in the gel are cultured in 24-well culture plates (Fig. 1a, b) (*see* **Note 4**).
7. These gel solutions are incubated in 37 °C CO_2 incubator for 10 min to form gel. Cholangiocyte culture medium (500 μL/well) is added, and cells are cultured for 10–12 days with medium changes every 3 days.
8. Cystic structures are formed in the extracellular matrix gel after 14–16 days of culture (Fig. 1c).

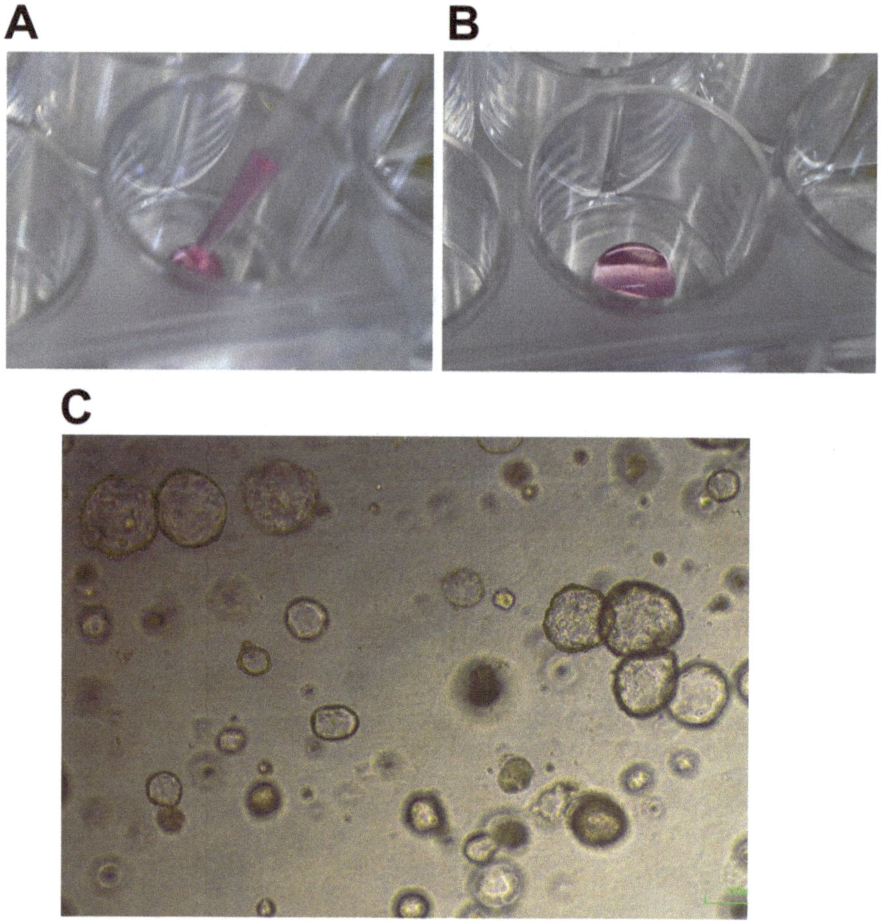

Fig. 1 Method for the culture of bile-duct like cysts in extracellular gels. (**a**) After mixing the cells into extracellular matrix gel solution, place the solution in the center of the well of the culture dish. (**b**) After incubation for 10 min in a CO_2 incubator, the gel solution formed gel. (**c**) Cyst structures derived from human iPS cells were formed

3.3 Characterization of Human iPS Cells-Derived Cholangiocytic Cells

3.3.1 Analyses of Cholangiocytic Differentiation Using RT-PCR

1. During differentiation steps (derived from Subheading 3.1), iPS cells-derived hepatic progenitor cells are washed with PBS, and total RNAs are extracted with RNAiso Plus (1 mL/well of 6-well culture dish). For the control, total RNA is purified from non-differentiated iPS cells. Individual extracellular matrix gels containing cholangiocytic cysts are lysed in RNAiso Plus (1 mL/well of 24-well culture dish). Total RNA is purified according to the manufacturer's protocol.
2. 0.5 μg Total RNA is used for single-strand cDNA synthesis using ReverTra Ace® qPCR RT Master Mix.
3. The expression levels of marker genes for cholangiocytic cells (Keratin 19, Keratin 7, and Grhl2) and hepatic cells (AFP and HNF4α) are examined in normal human iPS cells and differentiated iPS cell-derived cells. The expression of hypoxanthine

phosphoribosyltransferase 1 (HPRT1) is used as an internal control. RT PCR is performed using Roche Universal Library and THUNDERBIRD® Probe qPCR Mix according to the manufacturer's protocol. PCR primers and probes are shown in Table 2.

3.3.2 Analyses of Cholangiocytic Differentiation Using Immunocytochemistry

1. Individual extracellular matrix gels containing cholangiocytic cysts are washed with PBS and incubated with collagenase solution (500 μL/well of 24-well culture dish) for 15 min at 37 °C (*see* **Note 5**).
2. After washing with PBS, cells are fixed with 4% paraformaldehyde (500 μL/well) for 30 min at room temperature (RT).
3. After the washing step with PBS, cells are permeabilized with blocking medium (500 μL/well). Cells are incubated for 30 min at RT.
4. The cells are then incubated with diluted primary antibodies overnight at 4 °C. Primary antibodies are diluted with blocking medium.
5. After the washing step with blocking medium, cells are then incubated with diluted secondary antibodies diluted with blocking medium for 3 h at 37 °C. In addition, nuclei are stained with 1× DAPI solution.
6. After the washing step with blocking medium, individual extracellular matrix gels containing cholangiocytic cysts are put on cover glasses and analyzed under a LSM700 confocal microscope (Fig. 2). In order to avoid the dry of samples, we cover gel samples using immunofluorescent mount solution.

4 Notes

1. Feeder-free iPS cells are also useful for differentiation into hepatoblasts and cholangiocytes. We have succeeded cholangiocytic cyst formation derived from feeder-free human iPS cells using Cellartis® iPS Cell to Hepatocyte Differentiation System.
2. During the trypsinized step of cytokine-stimulated human iPS cells (Subheading 3.1), MEF feeder cells can also be collected. These contaminating cells are eliminated in the flow cytometry step.
3. To centrifuge a very small number of cells, use a specific tube (low cell adsorption tube STEMFULL™) that can block nonspecific binding to proteins, peptides, and cell surfaces.
4. Matrigel solution and collagen solution have to be kept cold in order to avoid unnecessary gel forming.

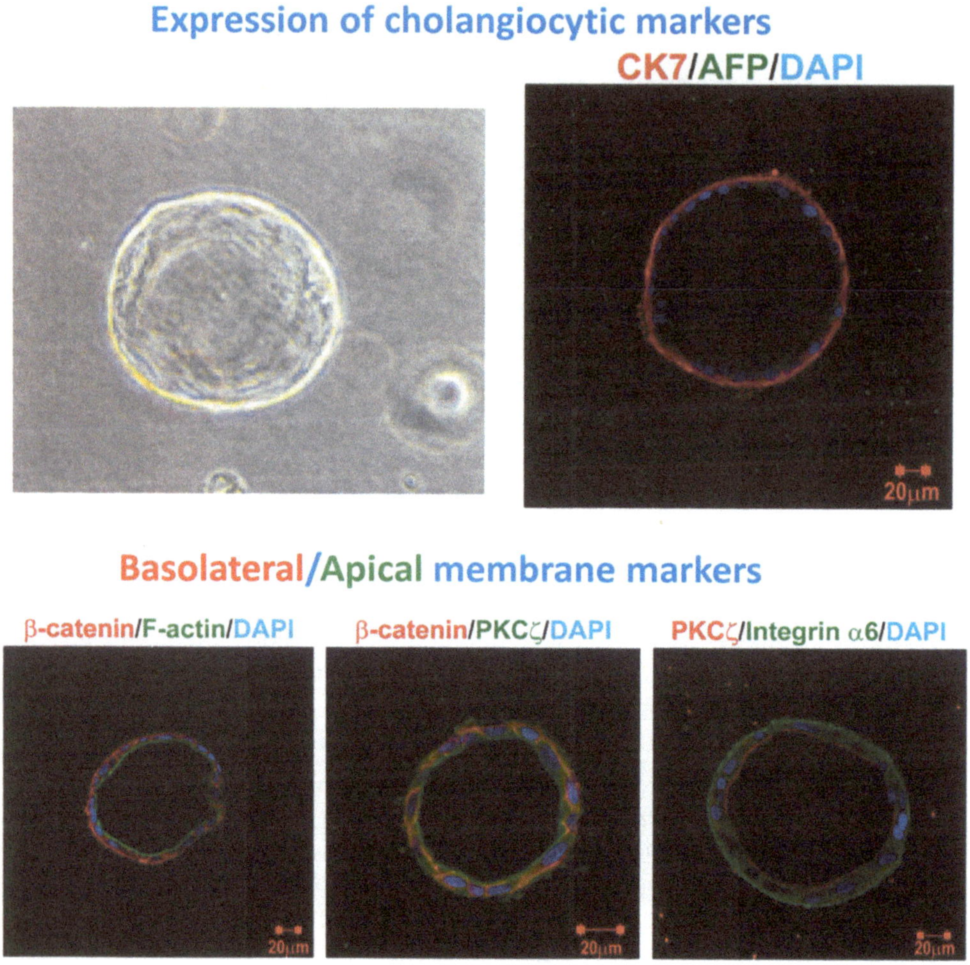

Fig. 2 Representative images of bile duct-like cysts derived from human iPS cells. Cysts were stained with antibodies against AFP, Keratin 7 (CK7), β-catenin, protein kinase Cζ, and integrin α6. F-actin was stained with phalloidin. Nuclei were counterstained with DAPI

5. During the staining steps, the gels containing cholangiocytic cysts are easily aspirated using the mechanical aspirator, which is usually used to remove supernatant. In order to avoid it, we remove supernatant using a micropipette.

Acknowledgments

This work was supported in part by the Education and Research Support Center of Tokai University. Some figures have been reproduced from [12]. This work was supported by a Grant-in-Aid for JSPS Fellows and a Grant-in-Aid from the Ministry of Education, Culture, Sports, Science and Technology, Japan.

References

1. Lemaigre FP (2009) Mechanisms of liver development: concepts for understanding liver disorders and design of novel therapies. Gastroenterology 137(1):62–79
2. Douarin NM (1975) An experimental analysis of liver development. Med Biol 53(6):427–455
3. Houssaint E (1980) Differentiation of the mouse hepatic primordium. I. An analysis of tissue interactions in hepatocyte differentiation. Cell Differ 9(5):269–279
4. Lemaigre F, Zaret KS (2004) Liver development update: new embryo models, cell lineage control, and morphogenesis. Curr Opin Genet Dev 14(5):582–590
5. Clotman F, Jacquemin P, Plumb-Rudewiez N, Pierreux CE, Van der Smissen P, Dietz HC, Courtoy PJ, Rousseau GG, Lemaigre FP (2005) Control of liver cell fate decision by a gradient of TGF beta signaling modulated by Onecut transcription factors. Genes Dev 19 (16):1849–1854
6. Lemaigre FP (2008) Notch signaling in bile duct development: new insights raise new questions. Hepatology 48(2):358–360
7. Zhang D, Gates KP, Barske L, Wang G, Lancman JJ, Zeng XI, Groff M, Wang K, Parsons MJ, Crump JG, Dong PDS (2017) Endoderm Jagged induces liver and pancreas duct lineage in zebrafish. Nat Commun 8(1):769
8. Oikawa T, Kamiya A, Kakinuma S, Zeniya M, Nishinakamura R, Tajiri H, Nakauchi H (2009) Sall4 regulates cell fate decision in fetal hepatic stem/progenitor cells. Gastroenterology 136(3):1000–1011
9. Tanimizu N, Kobayashi S, Ichinohe N, Mitaka T (2014) Downregulation of miR122 by grainyhead-like 2 restricts the hepatocytic differentiation potential of adult liver progenitor cells. Development 141(23):4448–4456
10. Takebe T, Sekine K, Enomura M, Koike H, Kimura M, Ogaeri T, Zhang RR, Ueno Y, Zheng YW, Koike N, Aoyama S, Adachi Y, Taniguchi H (2013) Vascularized and functional human liver from an iPSC-derived organ bud transplant. Nature 499 (7459):481–484
11. Koui Y, Kido T, Ito T, Oyama H, Chen SW, Katou Y, Shirahige K, Miyajima A (2017) An in vitro human liver model by iPSC-derived parenchymal and non-parenchymal cells. Stem Cell Rep 9(2):490–498
12. Yanagida A, Ito K, Chikada H, Nakauchi H, Kamiya A (2013) An in vitro expansion system for generation of human iPS cell-derived hepatic progenitor-like cells exhibiting a bipotent differentiation potential. PLoS One 8(7): e67541
13. Dianat N, Dubois-Pot-Schneider H, Steichen C, Desterke C, Leclerc P, Raveux A, Combettes L, Weber A, Corlu A, Dubart-Kupperschmitt A (2014) Generation of functional cholangiocyte-like cells from human pluripotent stem cells and HepaRG cells. Hepatology 60(2):700–714
14. Sampaziotis F, de Brito MC, Geti I, Bertero A, Hannan NR, Vallier L (2017) Directed differentiation of human induced pluripotent stem cells into functional cholangiocyte-like cells. Nat Protoc 12(4):814–827
15. Drenth JP, te Morsche RH, Smink R, Bonifacino JS, Jansen JB (2003) Germline mutations in PRKCSH are associated with autosomal dominant polycystic liver disease. Nat Genet 33(3):345–347
16. Davila S, Furu L, Gharavi AG, Tian X, Onoe T, Qian Q, Li A, Cai Y, Kamath PS, King BF, Azurmendi PJ, Tahvanainen P, Kaariainen H, Hockerstedt K, Devuyst O, Pirson Y, Martin RS, Lifton RP, Tahvanainen E, Torres VE, Somlo S (2004) Mutations in SEC63 cause autosomal dominant polycystic liver disease. Nat Genet 36(6):575–577
17. Wills ES, Te Morsche RHM, van Reeuwijk J, Horn N, Geomini I, van de Laarschot LFM, Mans DA, Ueffing M, Boldt K, Drenth JPH, Roepman R (2017) Liver cyst gene knockout in cholangiocytes inhibits cilium formation and Wnt signaling. Hum Mol Genet 26 (21):4190–4202
18. Itaba N, Wairagu PM, Aramaki N, Yasui T, Matsumi Y, Kono Y, Phan AN, Otsu M, Kunisada T, Nakamura Y, Okano H, Jeong Y, Shiota G (2014) Nuclear receptor gene alteration in human induced pluripotent stem cells with hepatic differentiation propensity. Hepatol Res 44(14):E408–E419
19. Yanagida A, Nakauchi H, Kamiya A (2016) Generation and in vitro expansion of hepatic progenitor cells from human iPS cells. Methods Mol Biol 1357:295–310

Part IV

Reconstitution of Liver Tissue Structures

Chapter 14

Generation of Hepatic Tissue Structures Using Multicellular Spheroid Culture

Fumiya Tao, Hirotaka Mihara, and Nobuhiko Kojima

Abstract

The hepatic functions of the hepatocytes in multicellular spheroid (MCS) are lower than those in the liver. One of the causes is that conventional hepatic MCSs do not reproduce liver-specific microstructures such as hepatic cord. It is necessary to design the inner structure of hepatic MCSs mimicking a structural feature of hepatic cord to represent further hepatic functions. Here we introduce a unique method to engineer the microarchitectures in the MCSs by formation of void spaces or filling of extracellular matrices (ECMs).

Key words Multicellular spheroid, Hepatocyte, Methylcellulose, Microstructure, Alginate hydrogel bead, Extracellular matrix

1 Introduction

Compared with monolayer cultures, MCSs are useful tools for mimicking a part of events that occur in biological tissues. However, the cellular microenvironments of hepatic MCSs are different from those of liver tissues. Liver tissues contain various microstructures like hepatic cords. Because these structures are considered to be responsible for the liver functions, it is desirable to reproduce the structures inside MCSs. Various methods for creating MCSs have been reported, as reviewed by Laschke et al. [1]. Nevertheless, all of these approaches do not generate complex liver tissue structures in MCSs. Therefore, it is necessary to develop a method that can design the inner structures of MCSs. We previously reported several cell aggregation methods using methylcellulose (MC) medium [2–7]. In this chapter, we introduce a unique method to gather hepatocytes and hydrogel beads by the MC medium. In addition, we describe a further application method of the MC medium in which ECMs are introduced into MCSs. These methods allow formation of MCSs with microchannel structures and/or thin layers of ECMs in a few days.

Naoki Tanimizu (ed.), *Hepatic Stem Cells: Methods and Protocols*, Methods in Molecular Biology, vol. 1905,
https://doi.org/10.1007/978-1-4939-8961-4_14,

2 Materials

All liquids and plasticware used for cell culture must be sterile. Regular culture medium is Dulbecco's Modified Eagle's Medium (DMEM) supplemented with 10% fetal bovine serum (FBS) and 100 U/mL penicillin/streptomycin. Phosphate-buffered saline (PBS) used uncontained calcium and magnesium ions.

2.1 Cell Cultures

Choose the appropriated culture medium and centrifugation conditions depending on the type of cells. In this chapter, we introduce MCS culture using human hepatoma cell line Hep G2 cells (*see* **Note 1**).

1. Polystyrene cell culture dish.
2. Trypsin-EDTA: 0.25 w/v% trypsin-1 mM ethylenediaminetetraacetic acid (EDTA).
3. Trypan blue solution.
4. Hemocytometer.

2.2 MCS Assembly

2.2.1 3% MC Medium

1. Methylcellulose (*see* **Note 2**).
2. Magnetic stirrer and magnetic bar.
3. 250 mL wide-mouthed glass bottle.
4. Cold room (4 °C) or 1 L beaker with crushed ice.

2.2.2 Alginate Hydrogel Beads

1. 1.5% sodium alginate solution: sodium alginate in Milli-Q. Sterilize the solution by autoclave.
2. 5% calcium chloride solution: calcium chloride in Milli-Q. Sterilize the solution by autoclave.
3. Inkjet system: WaveBuilder, PulseInjector (25 μm-diameter nozzle), and ink cartridge (*see* **Note 3**).
4. 1 mL syringe.
5. 40 μm cell strainer.
6. 0.22 μm syringe filter.
7. Magnetic stirrer and magnetic bar.
8. 100 mm petri dish.
9. 1000 μL micropipette.

2.2.3 Generation of MCS and Isolation of MCS from the MC Medium

1. Positive displacement pipette.
2. 35 mm petri dish.
3. 2 μL micropipette.
4. Medium containing diluted Matrigel: thaw appropriate amount of Matrigel tube overnight at 4 °C. Take the Matrigel tube from 4 °C and place them on ice. Then, Matrigel was

diluted to about 30 times by using FBS-free DMEM. Store it at 4 °C.

5. Ultralow attachment plate.
6. 5 U/mL cellulase solution: cellulase in regular culture medium.
7. 1000 μL micropipette with a truncated tip.
8. 1.5 mL micro tube.

3 Methods

3.1 3% MC Medium (See Note 4)

1. Put 3 g of MC powder in a wide-mouthed bottle with a magnetic bar, and sterilize it by autoclaving.
2. Add 100 mL of the regular culture medium to the wide-mouthed bottle.
3. Mix MC and regular culture medium with a magnetic stirrer at cold room or in a 1 L beaker with crushed ices overnight (*see* **Notes 5–7**).
4. Store the wide-mouthed bottle at 4 °C after MC is completely dispersed (*see* **Note 8**).

3.2 Production of Alginate Hydrogel Beads

Sodium alginate solution can turn into a gel after contacting with calcium ions. To produce alginate hydrogel beads, the inkjet system is used to discharge droplets of sodium alginate solution into calcium chloride solution.

1. Pour 5% calcium chloride solution in a 100 mm petri dish with a magnetic bar, and stir it at about 200 rpm.
2. Filter 1.5% sodium alginate solution using a 0.22 μm syringe filter to remove small dusts because those may be jammed in a PulseInjector, and fill an ink cartridge with sodium alginate solution using a 1 mL syringe.
3. Attach the ink cartridge to a PulseInjector. Discharge sodium alginate solution to calcium chloride solution at a voltage of 15 V and a frequency of 1000 Hz for several hours (*see* **Note 9**).
4. Collect alginate hydrogel beads with calcium chloride solution in a 50 mL tube using a 1000 μL micropipette, and filtrate it through a 40 μm cell strainer to remove particles with the diameter of 40 μm or more.
5. Centrifuge for 3 min at 200 × *g* and aspirate the supernatant.
6. Add 5 mL PBS to the tube and pipette up and down gently.
7. Repeat **step 5**.
8. Resuspend hydrogel beads with 5 mL regular culture medium, and take a small aliquot to a hemocytometer.

9. Count hydrogel beads.
10. Repeat **step 5**.
11. Adjust the concentration of hydrogel beads to 2×10^6 particles/mL by adding regular culture medium, and store it at 4 °C (*see* **Note 10**).

3.3 Generation of Normal MCS Using MC Medium

Figure 1 shows the schematic overview of the aggregation in the MC medium.

1. Culture Hep G2 cells in polystyrene cell culture dishes. Aspirate the medium and wash in PBS.
2. Remove PBS, and add appropriate volume of trypsin-EDTA for recovering cells to each dish. Incubate the dish at 37 °C for 3 min.
3. Add 10 mL regular culture medium to stop trypsinization.
4. Collect cells into a 15 mL centrifuge tube.

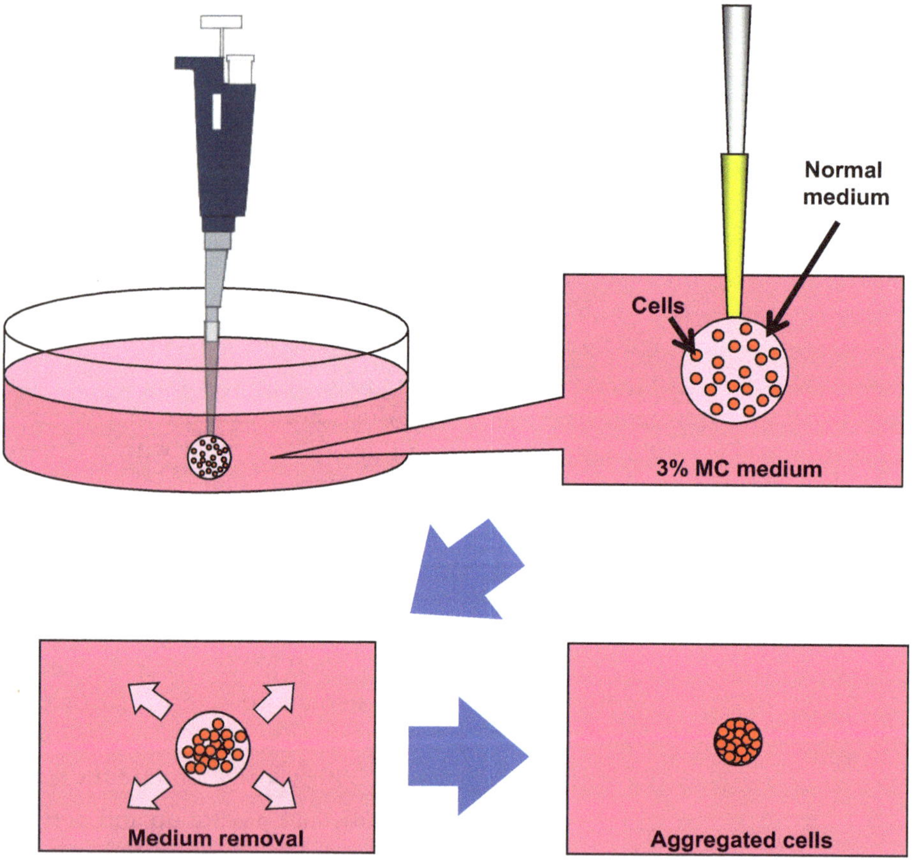

Fig. 1 Schematic model of the cell aggregation in the 3% MC medium. Cells are shown as red color. Because of its swelling property, the MC medium started absorbing the regular medium immediately. Finally, suspended cells were gathered and form a MCS in the MC medium

5. Centrifuge for 3 min at 200 × *g* and aspirate the supernatant.
6. Resuspend cells with 5 mL regular culture medium, and take a small aliquot for cell counting.
7. Mix an equal volume aliquot of both cell suspension and trypan blue solution, and count live and dead cells using a hemocytometer (*see* **Note 11**).
8. Repeat **step 5**.
9. Adjust the cell number to 2×10^6 cells/mL by adding regular culture medium.
10. Pour the MC medium into a petri dish using a positive displacement pipette (*see* **Notes 12** and **13**).
11. Inject 1 μL of the cell suspension into the MC medium using a micropipette (*see* **Note 14**).

3.4 Generation of MCS with Microchannel Structures

Continuously connected hydrogel beads in MCSs formed microchannel structures, which resembles microvasculature. The microchannel structures improved the albumin secretion rate and suppressed the expression of hypoxia-inducible factor-1α in relatively large (≧700 μm in diameter) MCSs composed of Hep G2 cells [4].

1. Follow Subheading 3.3, **steps 1–9** to obtain the cell suspension.
2. Mix equal volumes of the cell suspension and the hydrogel bead suspension.
3. Pour the MC medium into a petri dish using a positive displacement pipette.
4. Inject 1 μL of the mixture of cells and beads into the MC medium to assemble MCSs with microchannel structures (*see* **Note 15**) (Fig. 2).

3.5 Generation of MCS with Thin ECM Layer (See Note 16)

1. Follow Subheading 3.3, **steps 1–9** to obtain the cell suspension.
2. Centrifuge for 3 min at 200 × *g* and aspirate the supernatant.
3. Resuspend cells with medium containing diluted Matrigel on ice.
4. Pour the MC medium into a petri dish using a positive displacement pipette.
5. Inject 1 μL of cell suspension containing diluted Matrigel into the MC medium to assemble MCSs composed of cells and ECM (Fig. 3).

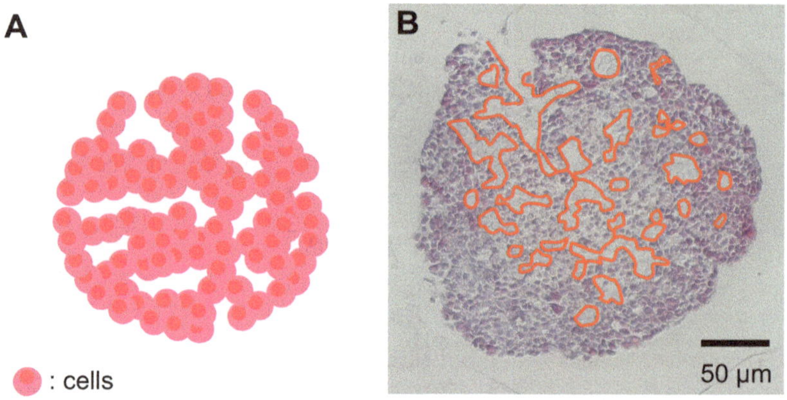

Fig. 2 MCSs with microchannel structures. (**a**) Schematic model of MCSs with microchannel structures. (**b**) The hematoxylin and eosin staining with sectioning of MCSs with microchannel structures. Red line demonstrates microchannel structures

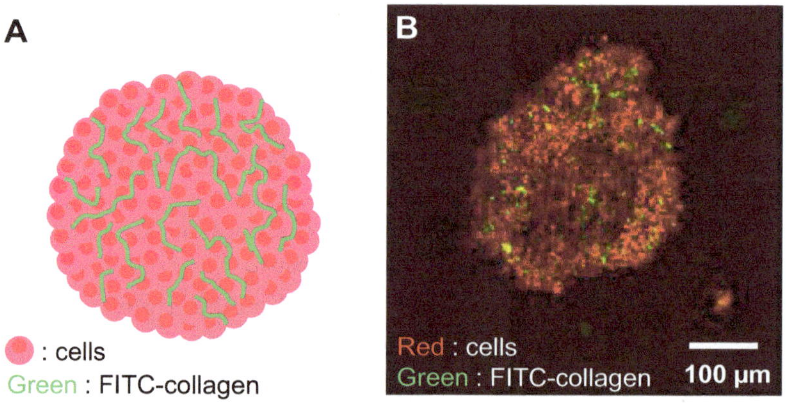

Fig. 3 MCSs with ECM thin layer. (**a**) Schematic model of MCSs with ECM thin layer. (**b**) Image of MCS with thin ECM layer obtained by confocal microscopy. This figure was used FITC labeled collagen instead of Matrigel to visualize ECM. Red color shows cells. Green color indicates FITC-labeled collagen

3.6 Isolation of MCS from the MC Medium and Further Culture

MCSs can be isolated from the MC medium after 1 day (*see* **Note 17**). Two isolation methods are described below. Choose an appropriate isolation method depending on the number of MCSs (*see* **Note 18**). Isolated MCSs can be further cultured by transferring MCSs to ultralow attachment plates (*see* **Note 19**).

3.6.1 Isolation of a Small Number of MCSs

1. Take a small amount of the MC medium containing MCSs using a micropipette with a truncated tip, and transfer MCSs to a 1.5 mL micro tube.
2. Add 1 mL PBS to remove the MC medium.
3. Wait for MCSs to sink to the bottom of the tube and aspirate the supernatant.
4. Add an appropriate volume of regular culture medium.

3.6.2 Isolation of a Large Number of MCSs

1. Overlay the 5 U/mL cellulase solution on the MC medium at 37 °C for about 30 min to reduce the viscosity of the medium (*see* **Notes 20–22**).
2. Transfer MCSs from the MC medium using a micropipette with a truncated tip to an adequate tube.
3. Wash MCSs using PBS to remove the MC medium.
4. Wait until MCSs sink to the bottom of the tube and aspirate the supernatant.
5. Add an appropriate volume of regular culture medium.

4 Notes

1. The 3% MC medium can generate MCSs using cells described below: primary human, mouse or rat hepatocytes, human iPSC-derived hepatocytes, mouse pancreatic β-cell line, primary mouse or rat pancreatic β-cells, primary mouse or rat bone marrow cells, primary human vascular endothelial cells, and primary mouse testicular cells.
2. Product number: M0512, Sigma-Aldrich is especially recommended.
3. WaveBuilder can adjust discharge conditions of a PulseInjector with three parameters (drive waveform, number of discharge per second, and drive voltage). PulseInjector discharges liquid particles. Ink cartridge is liquid container.
4. The concentration of the MC medium can be changed. Generally, the cells are less likely to be aggregated, as the concentration decreases. Maximum concentration of MC is 3% because it becomes hard to disperse completely.
5. Lower temperatures are better to disperse MC.
6. As MC is dispersed, the viscosity of medium is increased. Check the bottle occasionally to make sure the bar is continuously spinning.
7. A powerful magnetic stirrer is recommended.
8. The 3% MC medium is made by a day before use because 1 day is required to disperse MC. If there are masses of MC, keep mixing the medium with a magnetic stirrer. A few MC residues can be dispersed by leaving the bottle at 4 °C. The MC medium is stable when it is stored at 4 °C; however, it should be used as soon as possible.
9. Adjust the voltage and frequency in order to fly uniform droplets from the nozzle. Depending on temperature and humidity, optimum conditions of voltage and frequency are changed.

The strobe photography can be used to confirm whether droplets are uniform.

10. The density of hydrogel beads can be changed depending on the applications. The bead content is able to be changed from 0% to 60% in MCSs. When the bead content becomes greater than 60%, MCSs are fragile.
11. The live cells are not colored, and the dead cells appear as a blue color.
12. For example, 35 and 60 mm petri dishes are poured 2 and 4 mL of the MC medium, respectively.
13. When the MC medium is poured into a petri dish, a positive displacement pipette is used because the 3% MC medium is highly viscous. If bubbles are present in the MC medium, leave the dish until all the bubbles disappear.
14. Cells injected into the MC medium are rapidly gathered within the first 10 min. Although further incubation is effective, the movement becomes slow and almost stops within 30 min.
15. MCSs holding hydrogel beads have no liquid flow, but materials are exchanged between medium and MSCs in a diffusion manner because the beads are composed mostly of water. If material exchange by convection manner is desired, the beads can be removed by alginate lyase digestion to make void spaces. To digest beads, add culture medium containing 200 μg/mL alginate lyase at 37 °C for 5 min. Without the beads, however, MCSs are fragile. Choose stable/diffusion manner or unstable/convection manner depending on the applications.
16. The tips of the micropipette and the 3% MC medium are cooled down until the injection procedure to prevent unexpected gelation. Injection is performed at room temperature.
17. It is possible to culture for several days in the MC medium.
18. Cellulase treatment can be used to reduce the viscosity of the MC medium to allow handling of a large number of MCSs.
19. Adjust the number of MCSs to avoid the contact among them, which causes fusion of MCSs.
20. For example, when 2 mL the MC medium is used, the volume of cellulase solution is 1 mL.
21. Shake the dish of the MC medium occasionally.
22. When the viscosity is not sufficiently reduced in 30 min, keep the dishes to extend digestion at 37 °C.

Acknowledgment

This work was supported by JSPS KAKENHI under Grant Number 24710142 and AMED under Grant Number JP17be034302.

References

1. Laschke MW, Menger MD (2017) Life is 3D: boosting spheroid function for tissue engineering. Trends Biotechnol 35:133–144
2. Kojima N, Takeuchi S, Sakai Y (2012) Rapid aggregation of heterogeneous cells and multiple-sized microspheres in methylcellulose medium. Biomaterials 33:4508–4514
3. Kojima N, Takeuchi S, Sakai Y (2014) Engineering of pseudoislets: effect on insulin secretion activity by cell number, cell population, and microchannel networks. Transplant Proc 46:1161–1165
4. Kojima N, Takeuchi S, Sakai Y (2014) Fabrication of microchannel networks in multicellular spheroids. Sensor Actuat B Chem 198:249–254
5. Motoyama W, Sayo K, Mihara H et al (2016) Induction of hepatic tissues in multicellular spheroids composed of murine fetal hepatic cells and embedded hydrogel beads. Regen Ther 3:7–10
6. Kamitori S, Ozeki Y, Kojima N (2016) β-Galactoside-mediated tissue organization during islet reconstitution. Regen Ther 3:11–14
7. Sayo K, Aoki S, Kojima N (2016) Fabrication of bone marrow-like tissue in vitro from dispersed-state bone marrow cells. Regen Ther 3:32–37

Chapter 15

Reconstruction of Hepatic Tissue Structures Using Interstitial Flow in a Microfluidic Device

Ryo Sudo

Abstract

Construction of three-dimensional (3D) hepatic tissue structures is important for in vitro tissue engineering of the liver, because 3D culture of hepatocytes is critical for the maintenance of liver-specific functions. Although conventional 3D culture methods are useful for constructing 3D hepatic tissue structures, the precise control of culture microenvironments is required to construct more physiological tissues in vitro. Recent advances in microfluidics technologies have allowed us to utilize microfluidic devices for hepatic cell culture, which opened the door for creating more physiological 3D culture models of the liver. Here, we describe the method for the construction of hepatic tissue structures using a microfluidic device which has a 3D gel region with adjacent microchannels. Primary rat hepatocytes are seeded into a microchannel in a microfluidic device. The cells are then cultured in interstitial flow conditions, which leads to the construction of 3D tissue structures.

Key words Microfluidic device, Interstitial flow, 3D culture

1 Introduction

Construction of three-dimensional (3D) hepatic tissue structures is important for in vitro tissue engineering of the liver, because 3D culture of hepatocytes is critical for the maintenance of liver-specific functions. Various 3D culture methods have been described for the construction of liver tissues in vitro [1]. For example, the spheroid culture is one of the first attempts to culture hepatocytes in a 3D configuration. In this culture method, hepatocytes are seeded on the culture surface preventing attachment, which leads to the formation of multicellular aggregates known as hepatocyte spheroids. 3D stacked-up culture is another approach to construct 3D hepatic tissue structures. In this culture, two-dimensional (2D) cell layers of hepatocytes are stacked to form hepatic cord-like structures. Although these 3D culture methods are useful for the construction of 3D hepatic tissue structures, the precise control of culture

Naoki Tanimizu (ed.), *Hepatic Stem Cells: Methods and Protocols*, Methods in Molecular Biology, vol. 1905,
https://doi.org/10.1007/978-1-4939-8961-4_15,

microenvironments is required to construct more physiological tissues in vitro.

Recent advances in microfluidics technologies have allowed us to utilize microfluidic devices for cell culture. Compared to conventional culture devices, microfluidic devices have advantages in terms of the precise control of culture microenvironments [1–3]. For example, microfluidic devices provide flexibility in controlling biochemical and biomechanical factors such as growth factor gradients and stiffness of extracellular matrix in culture. In addition, the design of microchannels allows to investigate interactions between multiple cell types in close proximity. Furthermore, these devices allow real-time monitoring of cellular morphogenesis. Therefore, the use of microfluidic devices for hepatic cells opens the door for creating more physiological 3D culture models in liver tissue engineering.

Here, we describe the method for the construction of hepatic tissue structures using microfluidic devices. Specifically, a microfluidic device, which has a 3D gel region with adjacent microchannels, is used for 3D culture of hepatocytes. Suspension of primary rat hepatocytes was injected into a microchannel, and the cells are cultured in interstitial flow conditions. Hepatocytes construct 3D tissue structures in the interstitial flow conditions.

2 Materials

2.1 Fabrication of Microfluidic Devices

1. SYLGARD® 184 silicone elastomer base and the curing agent.
2. Plastic cup.
3. Glass rod.
4. Desiccator connected with a vacuum pump.
5. Prepare an SU-8 mold for desired patterning of microchannels using standard photolithography technique (*see* ref. [4]).
6. Oven.
7. Biopsy punch.
8. Scotch tape.
9. Plasma cleaner.
10. Culture dish.

2.2 Surface Treatment of Microchannels and Gel Formation in a Microfluidic Device

1. Poly-D-lysine (PDL) solution: 1 mg/mL solution in deionized water. Store at −20 °C.
2. Collagen solution: Prepare 20 μL of 10× PBS in a 1 mL tube on ice. Add deionized water, 0.5 N NaOH (*see* **Note 1**), and rat tail type I collagen solution (*see* **Note 2**) to the 1 mL tube on ice, which makes up to 200 μL. Mix the solution gently by pipetting. Store on ice until use.

3. Humid chamber: Pour deionized water into a clean container (e.g., empty micropipette tip box), and autoclave it. Place the autoclaved container in a 5% CO_2 incubator at 37 °C until use.
4. Culture medium: Standard culture medium for hepatocyte culture can be used (*see* ref. [5]). DMEM supplemented with 20 mM HEPES, 25 mM $NaHCO_3$, 30 mg/L L-proline, 0.5 mg/L insulin, 10^{-7} M dexamethasone, 10% FBS, 10 mM nicotinamide, 1 mM ascorbic acid 2-phosphate, 10 ng/mL EGF, and antibiotics.

3 Methods

3.1 Fabrication of Microfluidic Devices

1. Prepare SYLGARD® 184 silicone elastomer base and the curing agent in a 10:1 weight ratio in a disposable plastic cup.
2. Mix the solution, which is poly-dimethyl siloxane (PDMS) prepolymer, by a glass rod.
3. Place the PDMS prepolymer in a desiccator connected with a vacuum pump for removing air bubbles in the PDMS prepolymer.
4. Pour the PDMS prepolymer onto an SU-8 mold to the desired thickness (e.g., 7 mm), and place them in a desiccator again for removing air bubbles completely.
5. Place the mold in an oven at 65 °C for >3 h to cure the degassed prepolymer (*see* **Note 3**).
6. Cut out and detach the cured PDMS from the mold using a scalpel.
7. Trim the cured PDMS, and punch the trimmed PDMS devices to form inlets and outlets of microchannels using a biopsy punch.
8. Remove small particles from the surface of the PDMS devices using Scotch tape (*see* **Note 4**).
9. Place the PDMS devices in a glass beaker with deionized water and autoclave them.
10. Place the PDMS devices in a clean container (e.g., empty micropipette tip box), and autoclave them.
11. Dry the autoclaved PDMS devices in an oven at 65 °C.
12. Place the PDMS devices, and autoclaved coverslips in a plasma cleaner for plasma treatment in air (*see* **Note 5**).
13. Bond a PDMS device and a coverslip immediately after air plasma treatment (Fig. 1).
14. Place the bonded PDMS device and coverslip on a flat surface, and gently press the PDMS device to enhance adhesion between the PDMS device and a coverslip.

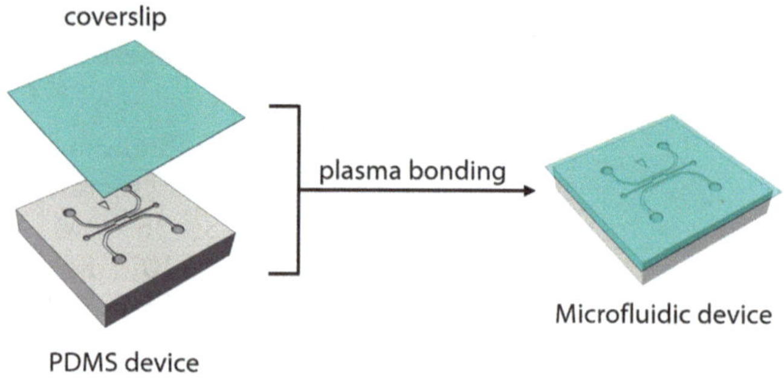

Fig. 1 Plasma bonding. A PDMS device and a coverslip are bonded after air plasma exposure, resulting in the fabrication of a microfluidic device

15. Place each microfluidic device in a culture dish after plasma bonding. The following procedures should be carried out aseptically.

3.2 Surface Treatment of Microchannels and Gel Formation in a Microfluidic Device

1. Fill channels of the microfluidic devices with PDL solution for surface coating (*see* **Note 6**).
2. Place the devices in a 5% CO_2 incubator at 37 °C for >4 h.
3. Remove the PDL solution and rinse the microchannels with deionized water twice.
4. Aspirate the microchannels completely and dry them in an oven at 65 °C overnight (*see* **Note 7**).
5. Prepare collagen solution on ice.
6. Slowly inject the collagen solution from an outlet of a gel channel to fill out the gel region using a 10 μL micropipette (Fig. 2a, b).
7. Place the devices in a humid chamber, and incubate them for 30 min in a 5% CO_2 incubator at 37 °C to allow gelation (*see* **Note 8**).
8. Gently fill the microchannels with warmed culture medium (Fig. 2c), and place the device in a culture dish.

3.3 Hepatocyte Seeding into a Microchannel

1. Prepare cell suspension of hepatocytes on ice, which can be isolated from rats by a two-step collagenase perfusion (*see* Chapter 5).
2. Add 20 μL of hepatocyte suspension at a density of 3×10^6 cells/mL to the outlet of a microchannel (Fig. 3a).
3. Tip the device to flow the cell suspension through the microchannel, which allows hepatocytes to attach and stack on the sidewall of collagen gel, resulting in the formation of multicellular aggregates (Fig. 3b, c) (*see* **Note 9**).

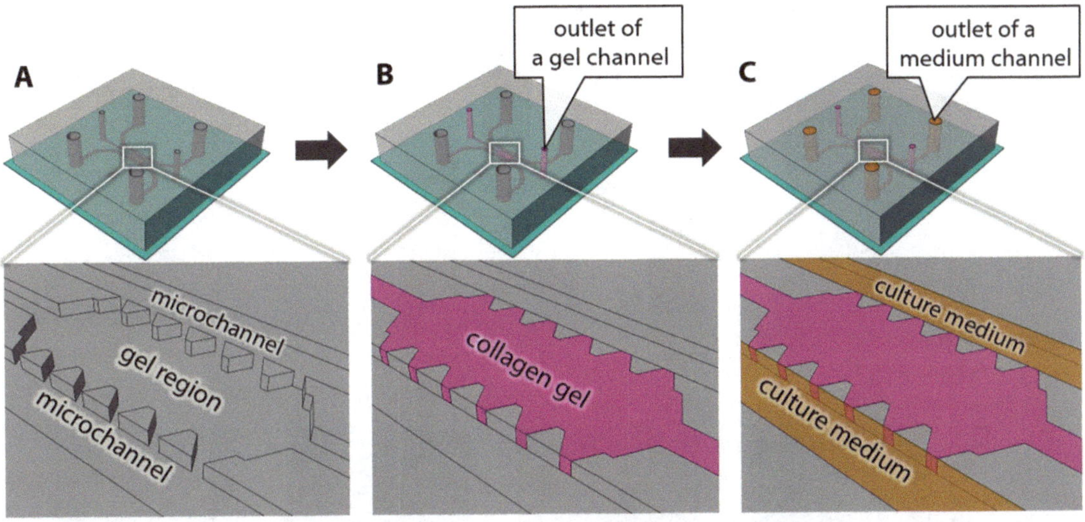

Fig. 2 Schematic images of a microfluidic device. (**a**) A microfluidic device before the injection of collagen solution. (**b**) A microfluidic device filled with collagen gel. Collagen solution is injected from an outlet of a gel channel to fill out the gel region. Gelation of the collagen solution results in the formation of two parallel microchannels for culture medium. (**c**) A microfluidic device filled with both collagen gel and culture medium. Microchannels are filled with culture medium after collagen solution is gelled

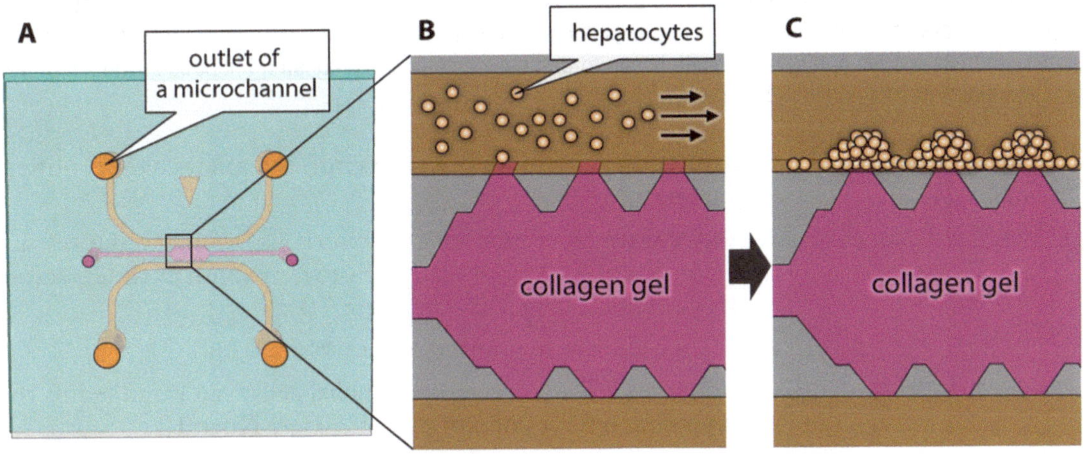

Fig. 3 Schematic images of a microfluidic device during hepatocyte seeding. (**a**) Cell suspension is added to the outlet of a microchannel. (**b**, **c**) A microchannel is filled with cell suspension, and the device is tipped on its side and maintained in the incubator to allow cells to attach and stack on the sidewall of collagen gel

4. Carefully place the devices in a humid chamber, and incubate them for 30 min in a 5% CO_2 incubator to allow cell adhesion.
5. Return the device to horizontal and place each device in a culture dish.

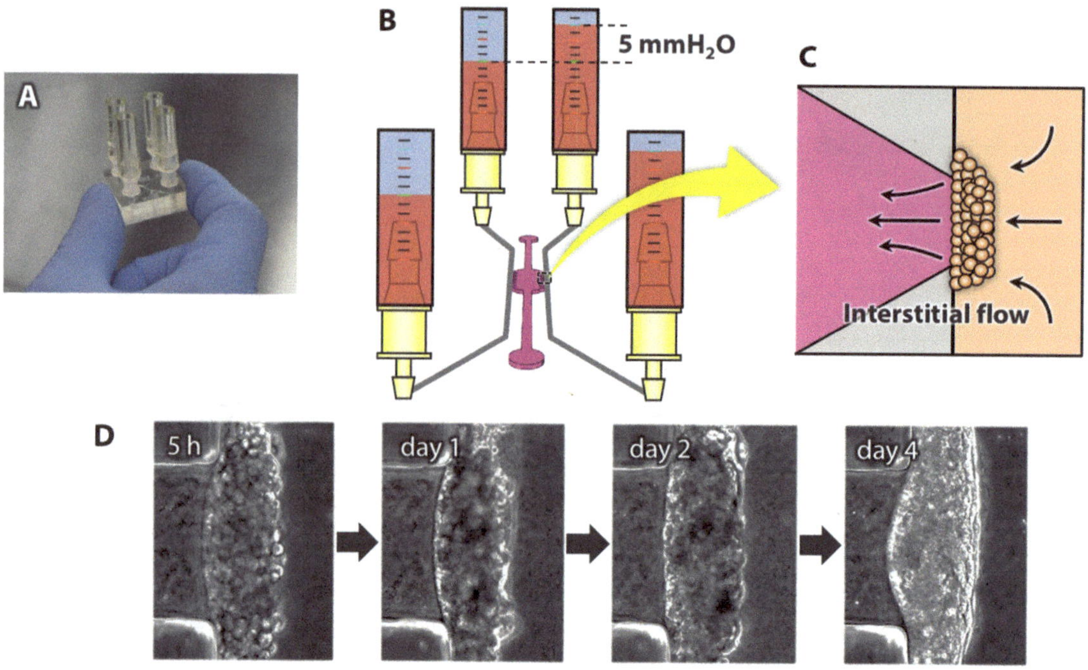

Fig. 4 Reconstruction of hepatic tissue structures in a microfluidic device. (**a**) A microfluidic device connected with medium reservoirs. (**b**) A schematic image of the microchannels with medium reservoirs. A pressure difference of 5 mmH_2O between two microchannels generates interstitial flow across the gel region. (**c**) Hepatocytes are stacked and construct multicellular aggregates under interstitial flow. (**d**) Corresponding phase-contrast images during hepatocyte tissue formation. Hepatocytes reconstruct hepatic tissue structures in interstitial flow conditions within 4 days of culture

3.4 Hepatocyte Culture in Flow Conditions

1. Gently insert medium reservoirs into the microchannel outlets (Fig. 4a).
2. Add culture medium to the reservoirs to make a pressure difference of 5 mmH_2O between two medium channels across the gel region (Fig. 4b) (*see* **Note 10**).
3. Change culture medium daily (*see* **Note 11**).
4. The other cell type (e.g., endothelial cells) can be added to the other channel for coculture (Fig. 5) (*see* **Note 12**).

4 Notes

1. The amount of 0.5 N NaOH is determined to adjust to the desired pH. We usually adjust pH 5–9. The pH during gelation is important for the stiffness of collagen gel, which results in different cellular morphogenesis in the gel (*see* refs. [6, 7]). For example, flexible collagen gel promotes sprouting angiogenesis while no significant effect of the gel stiffness on hepatocyte morphogensis was observed.

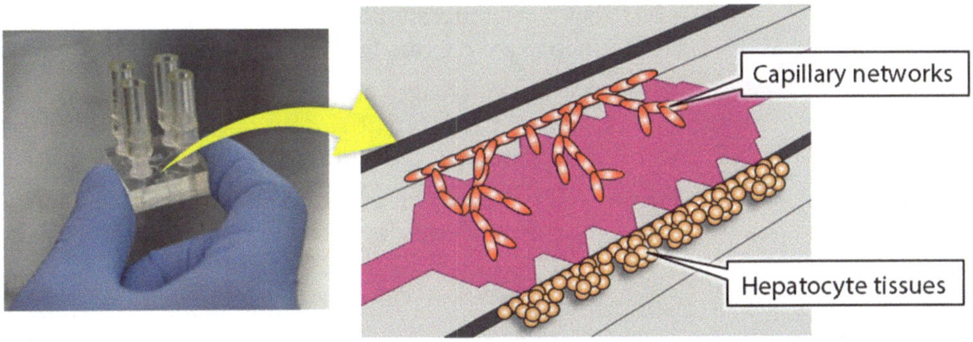

Fig. 5 An example of coculture in a microfluidic device. After hepatocytes construct 3D tissue structures, the other cell type (e.g., endothelial cells) can be added to the other channel

2. We use commercially obtained rat tail type I collagen solution, the concentration of which is ~4 mg/mL.
3. The temperature of the oven can be increased to shorten the curing time of PDMS.
4. This cleaning process should be carried out carefully since the remained small particles on the surface of the PDMS device disturb adhesion between the PDMS device and a coverslip during plasma bonding.
5. We usually perform the process of plasma exposure for ~60 s. However, this exposure time should be optimized depending on the specification of a plasma cleaner.
6. PDL coating of microchannels is carried out to avoid separation of collagen gel from channel walls. This process promotes 3D migration of cells in the collagen gel (*see* ref. [8]).
7. This dry process is important for restoring hydrophobicity of the channel surface. Hydrophobic channel surfaces are essential for retaining collagen solution in the desired position during gel formation described in Subheading 3.2, **step 6**.
8. The incubation time for gelation should be optimized depending on the volume of the gel. However, it is important to keep the incubation time to a minimum for reducing evaporation of collagen solution even in a humid chamber. Evaporation of collagen solution leads to insufficient gel volume.
9. The size of multicellular aggregates can be confirmed visually.
10. Hepatocytes form 3D clusters on the sidewall of collagen gel (Fig. 4c). These cells organize into hepatic tissue structures within 3–4 days of culture in interstitial flow conditions (Fig. 4d). We found that the flow direction from the cells to gel is important for the tissue formation. The pressure difference of 5 mmH_2O can be changed according to the desired velocity of interstitial flow. However, we confirmed that

interstitial flow generated by 5 mmH_2O is effective for inducing 3D hepatocyte tissue formation (*see* ref. [5]).

11. It is important to change culture medium periodically because the velocity of the interstitial flow decreases over time but is restored by changing the culture medium (*see* ref. [5]).
12. Various cell types can be added to the other channel for investigation of interactions between multiple cell types in close proximity.

Acknowledgment

This work was supported by JSPS KAKENHI Grant Number 16H03173 to R.S.

References

1. Sudo R (2014) Multiscale tissue engineering for liver reconstruction. Organogenesis 10:216–224
2. Chung S, Sudo R, Vickerman V, Zervantonakis IK, Kamm RD (2010) Microfluidic platforms for studies of angiogenesis, cell migration, and cell-cell interactions. Ann Biomed Eng 38:1164–1177
3. Zervantonakis IK, Kothapalli CR, Chung S, Sudo R, Kamm RD (2011) Microfluidic devices for studying heterotypic cell-cell interactions and tissue specimen cultures under controlled microenvironments. Biomicrofluidics 5:13406
4. Shin Y, Han S, Jeon JS, Yamamoto K, Zervantonakis IK, Sudo R, Kamm RD, Chung S (2012) Microfluidic assay for simultaneous culture of multiple cell types on surfaces or within hydrogels. Nat Protoc 7:1247–1259
5. Sudo R, Chung S, Zervantonakis IK, Vickerman V, Toshimitsu Y, Griffith LG, Kamm RD (2009) Transport-mediated angiogenesis in 3D epithelial coculture. FASEB J 23:2155–2164
6. Yamamura N, Sudo R, Ikeda M, Tanishita K (2007) Effects of the mechanical properties of collagen gel on the in vitro formation of microvessel networks by endothelial cells. Tissue Eng 13:1443–1453
7. Chung S, Sudo R, Mack PJ, Wan CR, Vickerman V, Kamm RD (2009) Cell migration into scaffolds under co-culture conditions in a microfluidic platform. Lab Chip 9:269–275
8. Chung S, Sudo R, Zervantonakis IK, Rimchala T, Kamm RD (2009) Surface treatment induced three dimensional capillary morphogenesis in a microfluidic platform. Adv Mater 21:4863–4867

Chapter 16

Generation of Hepatic Organoids with Biliary Structures

Takeshi Katsuda, Takahiro Ochiya, and Yasuyuki Sakai

Abstract

Incorporation of bile drainage system into engineered liver tissue is an important issue to advance liver regenerative medicine. Our group reported that three-dimensional (3D) coculture of fetal liver cells (FLCs) and adult rat biliary epithelial cells (BECs) allows reconstruction of hepatic spheroids that possess bile ductular structures. In this chapter, we describe the detailed protocol to isolate FLCs and BECs and to generate the spheroids with bile drainage system using these two types of primary cells.

Key words Biliary epithelial cell, Fetal liver cell, Bile ducts, Hetero-spheroid, 3D culture, Tissue engineering

1 Introduction

Bile drainage system is indispensable for preparing a transplantable liver tissue equivalent to serve as an alternative to a liver transplant. There is increasing evidence that heterotopic transplantation of liver tissue engineered with adult and fetal hepatocytes, as well as pluripotent stem cell-derived hepatic cells, is therapeutically beneficial [1–4]. In particular, coculture of hepatic cells with vascular endothelial cells has dramatically increased the in vivo functionality of the engineered hepatic tissue after transplantation [2, 3]. However, the scarcity of the biliary system questions the risk of cholestasis in the engineered hepatic tissue after its transplantation. Accordingly, researchers have recently attempted to generate biliary epithelial cells (BECs) from pluripotent stem cells for the potential use for regenerative medicine [5–7].

In our previous study, we investigated the feasibility of incorporation of biliary structures into in vitro engineered liver tissue [8]. This study demonstrated that adult BECs exhibited the capacity to reconstruct ductular and cystic structures with biliary functionality. Such biliary reconstructive capacity was much higher in adult BECs than in fetal liver cells (FLCs). This finding, thus, suggests the important role of committed BECs rather than liver

Naoki Tanimizu (ed.), *Hepatic Stem Cells: Methods and Protocols*, Methods in Molecular Biology, vol. 1905, https://doi.org/10.1007/978-1-4939-8961-4_16,

progenitor cells in reconstituting the functional biliary system in vitro. In this chapter, we describe a step-by-step protocol to reconstruct hepatic hetero-spheroids possessing biliary structures using rat FLCs and adult BECs.

2 Materials

2.1 Animal and Reagents

1. E17.5 pregnant Wistar rats.
2. Isoflurane (Pfizer).
3. E-MEM.
4. Pre-perfusion buffer: Dissolving the following reagents in 500 mL distilled water: 40 g NaCl, 2.0 g KCl, 0.39 g $NaH_2PO_4 \cdot 2H_2O$, 0.76 g $Na_2HPO_4 \cdot 12H_2O$, 4.5 g glucose, 1.9 g EGTA, 3.7 g EDTA, 0.0030 g phenol red, and 11.9 g HEPES in 500 mL distilled water to prepare 10× stock solution. Diluting 50 mL of 10× stock solution with 450 mL distilled water. After autoclave sterilization, add 2.35 mL of 7.5% $NaHCO_3$ solution and 2 mL of 1 N NaOH solution.
5. 7.5% $NaHCO_3$ solution: Dissolve 7.5 g of $NaHCO_3$ in 100 mL dH_2O. Filter with a 0.22 μm filter, and store at 4 °C until use. Use within 1 month.
6. 0.05% collagenase solution: Dissolve 9.6 g NaCl, 0.48 g KCl, 0.094 g $NaH_2PO_4 \cdot 2H_2O$, 0.18 g $Na_2HPO_4 \cdot 12H_2O$, 2.86 g HEPES, 0.888 g $CaCl_2 \cdot 2H_2O$, and 0.42 g $NaHCO_3$ in 1.2 L of dH_2O supplemented with 14.4 mL of 0.05% phenol red solution. Store this basal solution at 4 °C until use. For each perfusion, pour 400 mL into a beaker, and adjust pH to 7–7.4 with 5 N NaOH and/or 6 N HCl. Then, dissolve 0.02 g trypsin inhibitor and 0.2 g collagenase. After adjusting the pH to 7.6 with 5 N NaOH and/or 6 N HCl, filter with a 0.22 μm filter, and store at 4 °C until use. The prepared collagenase solution should be used within 1 week.
7. Bile duct wash medium: Leibovitz L-15 medium (Invitrogen) supplemented with 20 mM HEPES (Sigma-Aldrich), 1.1 g/L galactose (Sigma-Aldrich), 30 mg/L L-proline (Sigma-Aldrich), 0.1 μmol/L insulin (Sigma-Aldrich), 10^{-7} M dexamethasone (Sigma-Aldrich), and antibiotics (Invitrogen).
8. BEC digestion medium: Leibovitz L-15 medium (Invitrogen) supplemented with 400 U/mL of collagenase (Wako), 700 U/mL of hyaluronidase (Sigma-Aldrich), 10^{-7} M insulin, and 10^{-7} M dexamethasone. Filter with a 0.22 μm filter, and store at 4 °C until use. The prepared solution should be used within 1 week.

9. BEC culture medium: DMEM supplemented with insulin-transferrin-selenium supplement (Invitrogen), 10^{-7} M dexamethasone, 10 mM nicotinamide (Sigma-Aldrich), 1 mM ascorbic acid (Wako), 10 ng/mL epidermal growth factor (EGF; Sigma-Aldrich), 10 ng/mL of hepatocyte growth factor (HGF; PeproTech), 10% FBS, and antibiotic/antimycotic solution (100 U/mL penicillin, 100 mg/mL streptomycin, and 0.25 mg/ mL amphotericin B) (Invitrogen). Filter with a 0.22 μm filter, and store at 4 °C until use.
10. FLC digestion solution: PBS supplemented with 0.1% collagenase. Filter with a 0.22 μm filter, and store at 4 °C until use. The prepared solution should be used within 1 week.
11. FLC culture medium: Williams' medium E supplemented with 2 mM Glutamax (Invitrogen), 10^{-6} M DEX, 10 ng/mL of EGF, 20 ng/mL of HGF, 20 ng/mL of fibroblast growth factor (FGF)-1 (PeproTech), 20 ng/mL of FGF-4 (PeproTech), 10 ng/mL of oncostatin M (Wako), 10^{-7} M insulin, 10^{-8} M glucagon (Sigma-Aldrich), 0.5 mM ascorbic acid, 10 mM nicotinamide, antibiotic/antimycotic solution (Invitrogen), and 10% FBS.
12. 10% FBS-DMEM (used as BEC washing medium and FLC decantation medium).
13. Mouse anti-CK19 antibody (Novocastra) (used at 1:50 dilution for IHC).
14. Goat anti-CK18 antibody (Santacruz) (used at 1:200 dilution for IHC).
15. 4% paraformaldehyde (PFA) (Wako).

2.2 Equipment

1. Peristaltic pump (EYELA, Tokyo, Japan, catalog number, RP-1000).
2. Silicon tube (Ø4.76 × 7.94 mm^2) (EYELA, catalog number, 125540).
3. Cannula Ø1.2 mm (plastic outer needle equipped with intravenous cannula (Top Corporation, Tokyo, Japan).
4. Incubator shaker (Bioshaker VBR-36 (Taitech)): An equivalent machine can be used.
5. Autoclaved 100 mL glass conical flask.
6. 40 μm cell strainer (BD).
7. 70 μm cell strainer (BD).
8. 25 mL reservoir (Nunc).
9. Eight-channel pipette (volume range, 10–50 μL, 30–300 μL) (Thermo).

10. 96-well, round-bottom plate coated with 2-methacryloyloxyethyl phosphorylcholine (MPC) (NOF CORPORATION).

3 Methods

3.1 Extraction of the Biliary Trees and Isolation of BECs

This procedure is performed in parallel with fetal liver extraction (*see* Subheading 3.2) using one pregnant rat.

1. Anaesthetize a pregnant rat by inhalation of isoflurane vapor. Place a paper towel cushion on a foam board or cork board, and lay the rat on the cushion. Sterilize the animal by wiping its fur with 70% EtOH. Open the abdomen using surgical scissors.
2. Start drip of pre-perfusion buffer at 25–30 mL/min. Half-cut the portal vein at the location approximately 1.5 cm from the bifurcation of the portal vein, and insert the cannula.
3. Immediately after confirming that the color of the liver changes from reddish brown to yellowish, cut the inferior vena cava. Perfuse approximately 450 mL pre-perfusion buffer (approximately 15 min).
4. Pause the pump, transfer the inlet tube to the collagenase bottle, and resume perfusion. Perfuse approximately 300–350 mL collagenase (approximately 10 min).
5. Stop the pump, and remove the cannula from the portal vein.
6. Peel the liver capsule, and gently mash the parenchyma with round-edged tweezers to release the parenchymal cells from the tree structures. Gently flush away the digested parenchymal tissue with E-MEM until white tree structures are visible (Fig. 1a–e).
7. Resect the tree structures, and place them in bile duct wash medium. At this moment, the tree structures still hold parenchymal cells as noticed by their brownish appearance.
8. Purify the tree structure ex vivo by further removing the remaining parenchyma until most brownish-colored parts are removed (Fig. 1f). Gently pressing the tissue with round-edged tweezers helps releasing of parenchymal cells from the tree structures.
9. Collect the purified tree fragments into multiple 1.5 mL tubes so that each tube contain 150–300 μL of fragments (usually divided into two to four tubes) (*see* **Note 1**), and mince them in the 1.5 mL tubes with surgical scissors into small fragments (approximately <0.5 mm). From this step, work aseptically in a clean bench or a safety cabinet.
10. Suspend the minced fragments in 10 mL BEC digestion medium, and divide it to two 50 mL tubes.

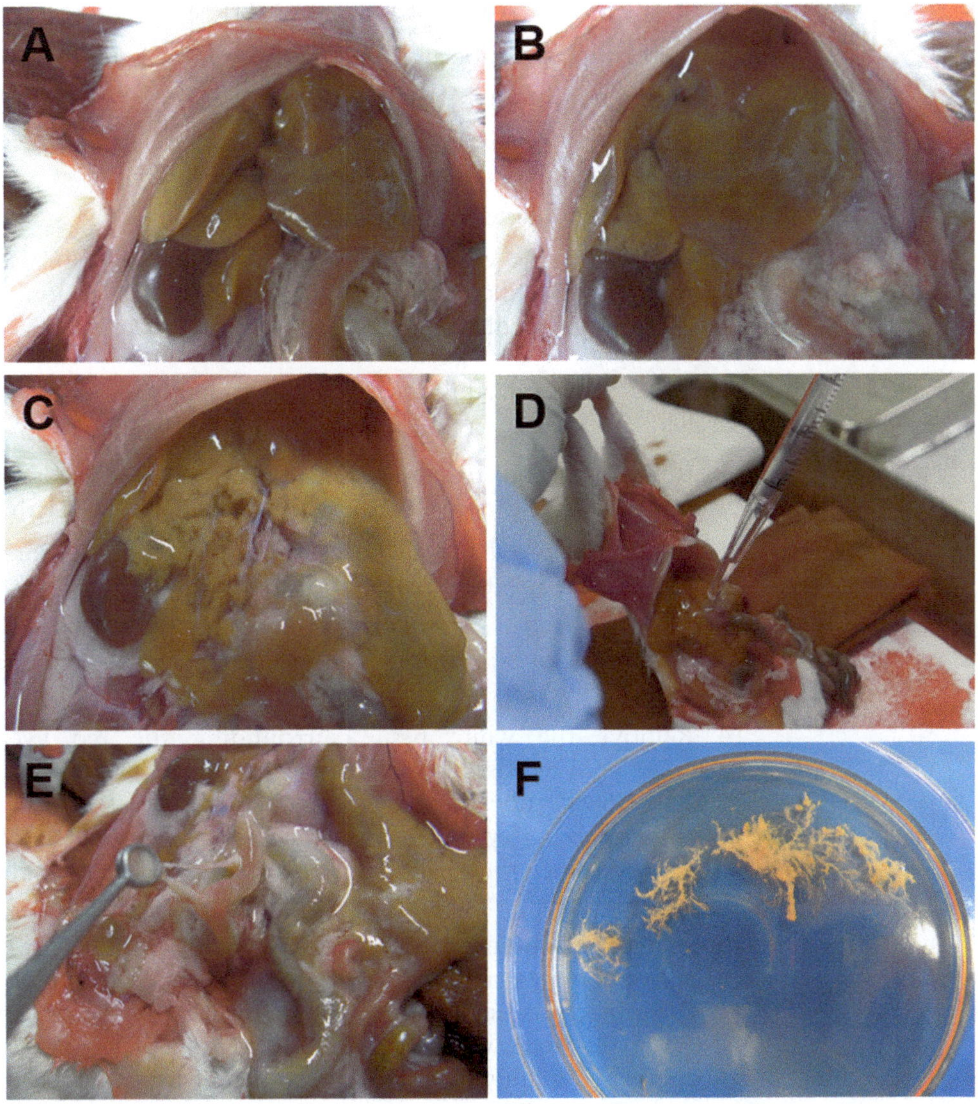

Fig. 1 Graphical schematic of extraction of intrahepatic bile duct trees. (**a**) Just after switching from pre-perfusion buffer to collagenase solution and (**b**) after completion of collagenase perfusion. Note that the liver is loosened and the subcapsular tissue is pulpy compared to (**a**). (**c**) By peeling the liver capsule and gently mashing the parenchyma with round-edged tweezers, parenchymal tissue can be liberated from the tree structure. (**d**) Flushing the parenchymal tissue with E-MEM by pipetting. Lifting up the rat's front paws will help this procedure. (**e**) White intrahepatic biliary trees can be visible after removal of the parenchymal tissue. (**f**) The purified tree structures after mechanically removing the remaining parenchyma using tweezers ex vivo

11. Shake the tubes at 37 °C in a shaking incubator for 30 min while making sure that the content is homogeneously mixed.
12. Filtrate the digested tissue with a 40 μm cell strainer, and combine dispersed cells from two tubes into a new 50 mL tube.
13. Centrifuge the tube at 800 × g for 10 min at 4 °C.

14. Resuspend in 10 mL of 10% FBS-DMEM.
15. Centrifuge the tube at 800 × *g* for 5 min at 4 °C.
16. Resuspend in 10 mL of 10% FBS-DMEM.
17. Count the approximate number of viable cells using a hemocytometer and trypan blue solution (*see* **Note 2**).
18. Centrifuge the tube at 800 × *g* for 5 min at 4 °C.
19. Resuspend the cells in appropriate volume of BEC culture medium to obtain 0.5–2 × 10^6 cells/mL suspension. Keep on ice until seeding.

3.2 Extraction of Fetal Livers and Purification of FLCs

While perfusing the adult liver, proceed to fetal liver extraction in parallel.

1. Remove the uterus containing fetuses from the body of the mother rat, and transfer them to a 10 cm dish filled with PBS.
2. Extract fetuses from the uterus using surgical scissors and tweezers, and transfer them to a new 10 cm dish filled with PBS.
3. Extract the livers from the fetuses using tweezers, and transfer the fetal livers to a new 10 cm dish filled with PBS.
4. Remove the white tissue surrounding the red livers as completely as possible using tweezers. Divide the purified livers into 1.5 mL tubes so that each tube contains 150–300 μL (two to five tubes, dependent on the number of fetuses) (*see* **Note 1**), and mince them with surgical scissors into small fragments (approximately <0.5 mm).
5. Transfer the minced tissue into a 15 mL tube, add PBS to adjust the total volume to 12 mL, and stand the tube until the minced tissues fall to the bottom (3–5 min) (*see* **Note 3**). From this step, work aseptically in a clean bench or a safety cabinet. Carefully discard the supernatant using a pipette, and add 12 mL PBS.
6. Repeat the wash process in **step 5** until the color of the supernatant changes from red to clear (usually two to three time repeats are sufficient).
7. Suspend the minced tissue in 2.5 mL/tube PBS, and add 2.5 mL of 0.1% collagenase solution (final concentration: 0.05%).
8. Shake the tube at 37 °C in a shaking incubator for 10 min. Make sure that the content is homogeneously stirred.
9. Let the tube stand for approximately 10 s, and retrieve the supernatant into a new tube while leaving the remaining fragments (*see* **Note 4**).
10. Centrifuge the tube at 200 × *g* for 2 min at 4 °C.

11. Discard the supernatant, and resuspend the cells in 10 mL of 10% FBS-DMEM.
12. Filter the cells with a 70 μm cell strainer into a new 100 mL tube. Wash the cell strainer by flushing it with fresh 90 mL 10% FBS-DMEM.
13. Decant the 100 mL cell suspension into an autoclaved glass conical flask. Let the flask stand at room temperature for 15 min. This decantation step helps removal of blood cells.
14. Carefully discard the upper 70 mL of the suspension, in which blood cells are enriched. Collect the bottom 30 mL suspension which is rich in FLCs into a new tube.
15. Count the viable cell number using a hemocytometer and trypan blue solution.
16. Centrifuge the tube at 200 × *g* for 2 min at 4 °C.
17. Resuspend the cells in appropriate volume of FLC culture medium to obtain 0.5–2 × 10^6 cells/mL suspension. Keep on ice until seeding.

3.3 Formation of BEC-FLC Hetero-spheroids

To obtain well-developed ductular structures, 50–75% of the mixed cells should be BECs [8]. Recommended seeding density is 1–20 × 10^4 cells/well of 96-well plate. In our previous report, we generated hetero-spheroids by seeding 2 × 10^5 cells/well in total, and we observed formation of ductular networks (*see* **Note 5**).

1. Pipette each cell suspension, and combine BECs and FLCs in a new tube at a desired mixing ratio.
2. Fill a 25 mL reagent reservoir with the mixed cell suspension. Seed to a round-bottom MPC-coated 96-well plate using an eight-channel pipette (*see* **Note 6**). Culture the cells at 37 °C in a CO_2 incubator. Thereafter, replace the medium with fresh FLC culture medium every day as described below. To confirm that BECs and FLCs are successfully harvested, it is recommended to seed a small portion of remaining cells to collagen-coated plates, and observe the morphologies of each cell type (Fig. 2).
3. On days 1 and 2, BECs and FLCs generate loose pancake-shaped aggregates. Since they are still very fragile, carefully replace the medium: remove approximately 70 μL of the culture supernatant, and add 100 μL fresh FLC culture medium into each well slowly not to mix up the cells.
4. Within 3 days of culture, the loose aggregates are compressed to form compact spheroids (Fig. 3). On day 3, remove cells that do not contribute to spheroid formation: stir the medium by pipetting with P200 tips (*see* **Note 7**), let the spheroids fall to the bottom, and remove the supernatant containing cells which have not contributed to spheroid formation.

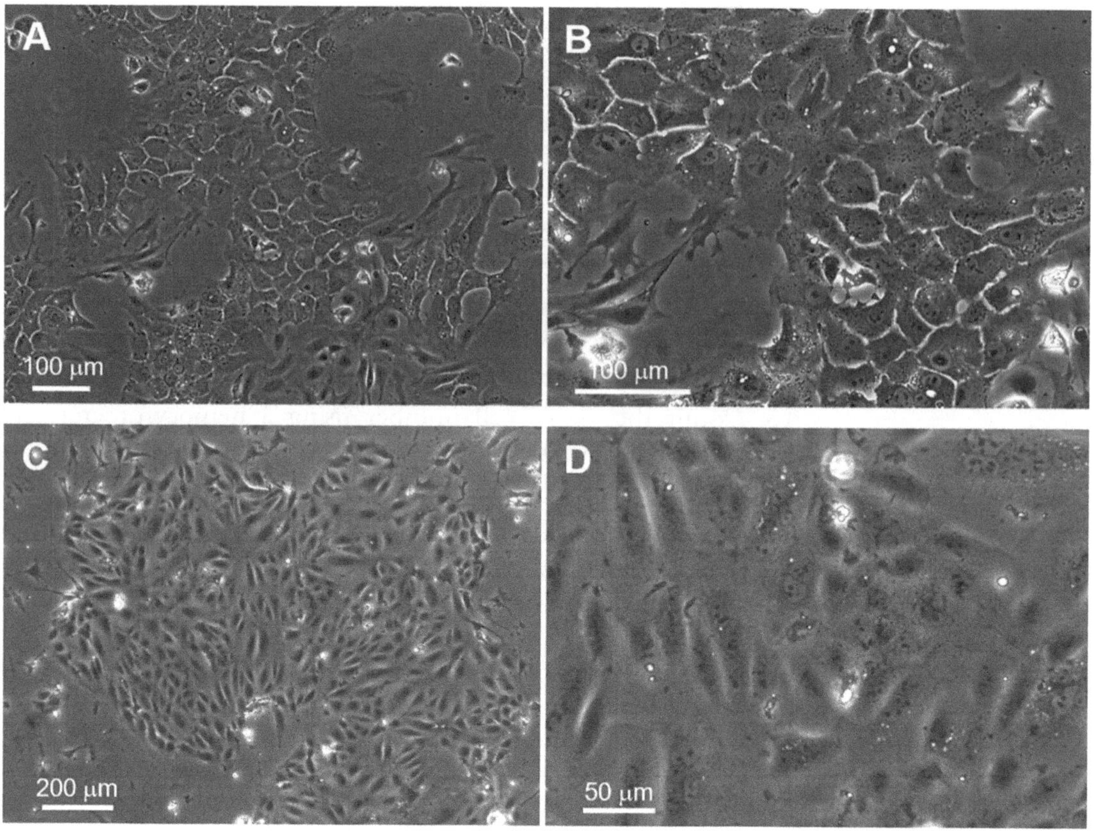

Fig. 2 Morphology of FLCs and BECs. (**a**, **b**) Phase-contrast images of E17.5 FLCs on day 2 of culture. (**c**, **d**) Phase-contrast images of BECs isolated from the adult rat liver on day 3 of culture. This figure is reproduced from ref. [8]

5. On days 3–6, fix the spheroids in 4% PFA, embed them in paraffin blocks, and prepare sections with approximately 5 μm thickness according to the general method (*see* **Note 8**). In the reconstructed hetero-spheroids, $CK18^{+}/CK19^{+}$ biliary structures develop mainly inside the spheroids, while $CK18^{+}/CK19^{-}$ hepatic structures are located mainly at outer regions (Fig. 4) (*see* **Note 9**).

4 Notes

1. Too much volume in a 1.5 mL tube disturbs efficient mincing.
2. Because it is difficult to completely dissociate biliary fragment into single cells, it is impossible to determine the precise cell number.

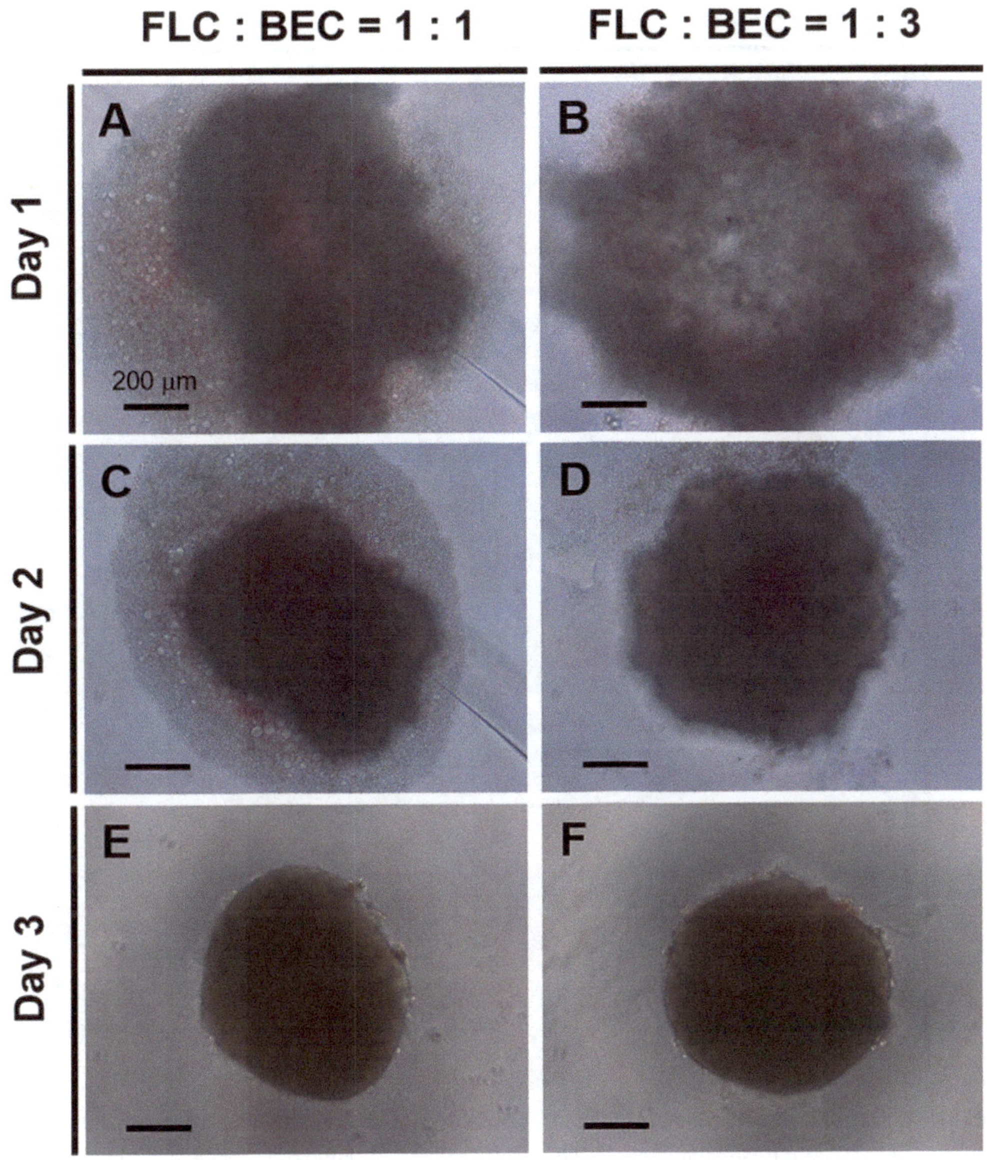

Fig. 3 Time course of hetero-spheroid formation. Phase-contrast images of hetero-spheroids with the composition of "FLC:BEC = 1:1" and "FLC:BEC = 1:3" hetero-spheroids on day 1 (**a**, **b**), day 2 (**c**, **d**), and day 3 (**e**, **f**). This figure is reproduced from ref. [8]

3. If the minced tissue volume is larger than 500 μL, aliquot the suspension to multiple tubes so that each tube contains less than 500 μL.
4. White tissue will be visible after digestion. Do not let the tube (s) stand too long to avoid loss of FLCs.
5. At the cell density of 2×10^5 cells/well, we observed necrosis at the central area of the spheroids, most likely because each spheroid contains too much cells. We suppose that a lower cell density

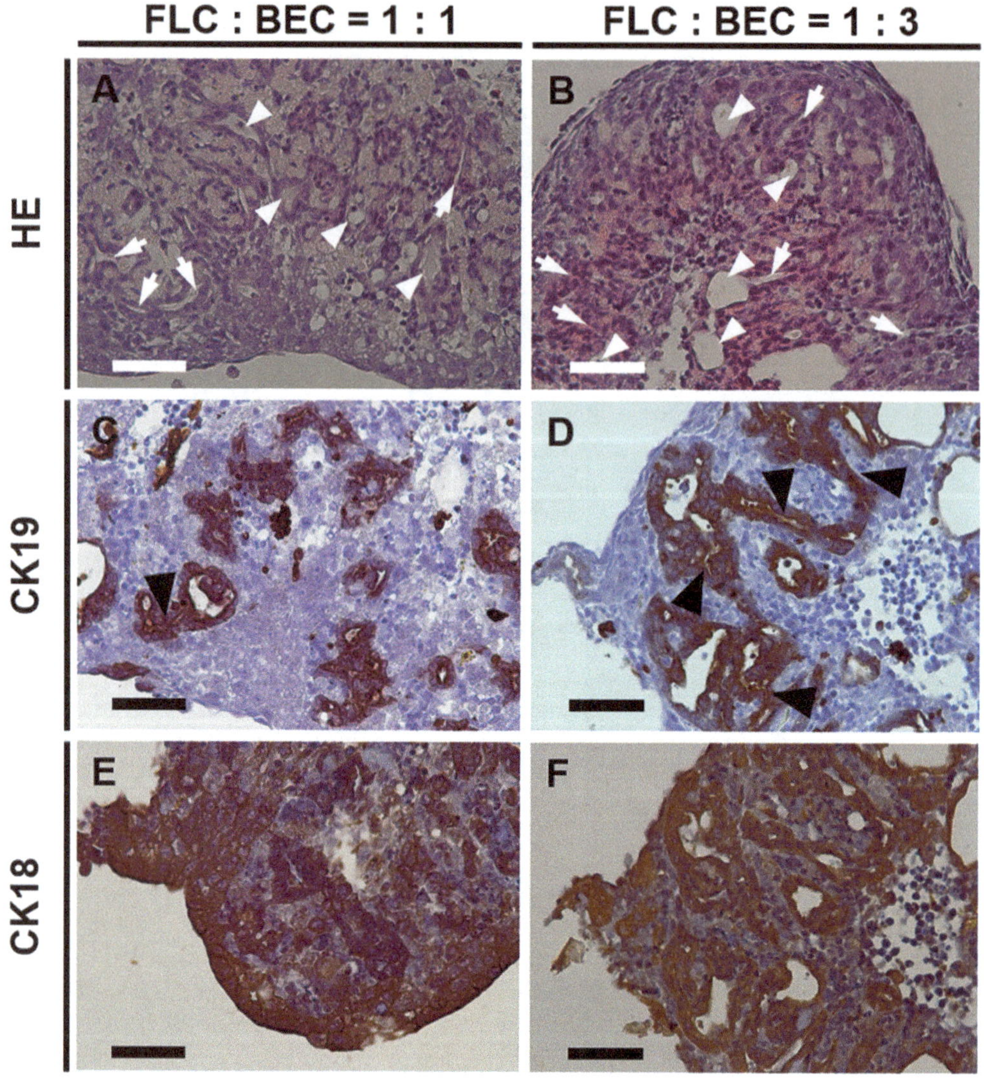

Fig. 4 Histological analyses of hetero-spheroids on day 3. (**a**, **b**) HE staining (nuclei, blue; cytoplasm, red-purple). Arrows and arrowheads indicate duct- and cyst-like structures, respectively. (**c**, **d**) IHC of CK19 (brown). Nuclei were counterstained with hematoxylin (blue). Arrowheads indicate continuous ductular networks formed by CK19 (+) BECs. (**e**, **f**) IHC of CK18 (brown). Nuclei were counterstained with hematoxylin (blue). Scale bar: 50 μm. *CK* cytokeratin, *HE* hematoxylin-eosin, *IHC* immunohistochemistry. This figure is reproduced from ref. [8]

would be better, although we have not yet optimized the number of seeding cells.

6. To avoid letting the cells fall to the bottom, seed cells as soon as possible.
7. Be careful not to break down the spheroids by touching them with the P200 tips.

8. When handing spheroids, we recommend to use nylon biopsy bags to reduce the risk of small specimen loss.
9. When performing immunohistochemistry, heat-induced antigen retrieval should be performed before antibody reaction. We usually heat de-paraffinized specimens in 1/200 diluted ImmunoSaver (Nissin EM Ltd.) at 98 °C using an electronic pot for 45 min.

Acknowledgments

This work was supported in part by Grant-in-Aid for Young Scientists B (16K16643) and the Research Program on Hepatitis (16fk0310505h0005) from the Japan Agency for Medical Research and Development (AMED).

References

1. Ohashi K, Yokoyama T, Yamato M et al (2007) Engineering functional two- and three-dimensional liver systems in vivo using hepatic tissue sheets. Nat Med 13:880–885. https://doi.org/10.1038/nm1576
2. Takebe T, Sekine K, Enomura M et al (2013) Vascularized and functional human liver from an iPSC-derived organ bud transplant. Nature 499:481–484. https://doi.org/10.1038/nature12271
3. Sakai Y, Yamanouchi K, Ohashi K et al (2015) Vascularized subcutaneous human liver tissue from engineered hepatocyte/fibroblast sheets in mice. Biomaterials 65:66–75. https://doi.org/10.1016/j.biomaterials.2015.06.046
4. Jiang J, Kojima N, Guo L et al (2004) Efficacy of engineered liver tissue based on poly-L-lactic acid scaffolds and fetal mouse liver cells cultured with oncostatin M, nicotinamide, and dimethyl sulfoxide. Tissue Eng 10:1577–1586. https://doi.org/10.1089/ten.2004.10.1577
5. Tian L, Deshmukh A, Ye Z, Jang YY (2016) Efficient and controlled generation of 2D and 3D bile duct tissue from human pluripotent stem cell-derived spheroids. Stem Cell Rev Rep 12:500–508. https://doi.org/10.1007/s12015-016-9657-5
6. Sampaziotis F, De Brito MC, Madrigal P et al (2015) Cholangiocytes derived from human induced pluripotent stem cells for disease modeling and drug validation. Nat Biotechnol 33:845–852. https://doi.org/10.1038/nbt.3275
7. Dianat N, Dubois-Pot-Schneider H, Steichen C et al (2014) Generation of functional cholangiocyte-like cells from human pluripotent stem cells and HepaRG cells. Hepatology 60:700–714. https://doi.org/10.1002/hep.27165
8. Katsuda T, Kojima N, Ochiya T, Sakai Y (2013) Biliary epithelial cells play an essential role in the reconstruction of hepatic tissue with a functional bile ductular network. Tissue Eng Part A 19:2402–2411. https://doi.org/10.1089/ten.TEA.2013.0021

Chapter 17

Analysis for Remodeling of Hepatic Tissue Structures in 3D During Regeneration

Kota Kaneko

Abstract

It has been demonstrated that the liver remarkably alter its tissue structures during regeneration, and various types of liver stem or progenitor cells locating in specific areas in the liver tissue contribute to regeneration. Therefore, it is important to analyze the dynamic rearrangement of the liver tissue structures in 3D for better understanding the process and mechanism of liver regeneration. Here we describe a macroscopic analysis method to visualize the whole 3D structure of the vasculatures and the biliary tree, which are dynamically remodeled during regeneration, and a microscopic analysis method to visualize detailed structures at the cellular level, which can compensate what cannot be detected in the macroscopic analysis.

Key words Liver regeneration, Liver pathology, 3D tissue structure, 3D imaging, Ductular reaction, Biliary tree

1 Introduction

Accumulating evidences have shown that the liver regeneration is supported by various types of stem or progenitor cells locating differentially in the tissue structure. For example, residential liver stem/progenitor cells are thought to be located near or inside the bile duct epithelium [1]. These cells are considered to contribute to regeneration, when the proliferative ability of mature hepatocytes is compromised. On the other hand, it has been suggested that pericentral or periportal hepatocytes have more regenerative potential than others [2–4]. Because of the diverse locations of these progenitors, it is necessary to analyze the regenerative processes correlating with the tissue structure. Moreover, since the liver tissue is organized in three dimension (3D), 3D analyses rather than histological analysis with 2D sections are needed for identifying the spatial localization of liver progenitor cells.

Although the liver is one of the largest organs, it has a repetitive structure called lobule. The lobules are defined by portal triads and

Naoki Tanimizu (ed.), *Hepatic Stem Cells: Methods and Protocols*, Methods in Molecular Biology, vol. 1905, https://doi.org/10.1007/978-1-4939-8961-4_17, © Springer Science+Business Media, LLC, part of Springer Nature 2019

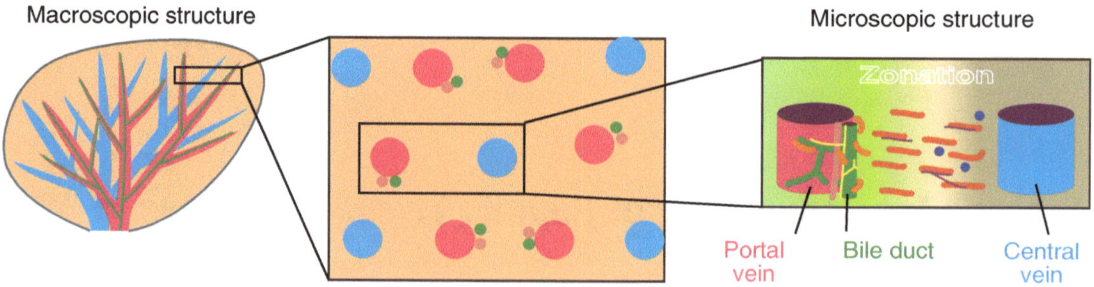

Fig. 1 The macroscopic and the microscopic structure of the liver. Bile ducts (green), portal vein (magenta), central vein (blue), sinusoids (red), hepatic artery (pink), neurons (yellow), and other immune and mesenchymal populations (orange, dark blue, and purple) are shown

the central veins, which are the landmarks to tell the "zonation." Therefore the whole liver structure can roughly be regarded as the collections of the lobules defined by vasculatures (Fig. 1). Around the portal vein exist bile ducts, hepatic artery, and the peripheral nerve systems. Importantly, bile ducts dynamically change the structure during liver injury and regeneration, which is called ductular reaction [5]. Recent studies have shown that bile ducts can extend their branches as far as to the pericentral area in response to injury, independently of the vascular structure [6–8]. Together with the classic view of the continuous epithelial structure of hepatocytes and biliary epithelial cells [1] (Fig. 2), such remodeling of the bile duct structure needs to be taken into consideration when assessing any spatial relationships. We recommend using portal veins, central veins, and bile ducts as basic landmarks to determine the locations of objects in the liver tissue.

There are many different methods to analyze the 3D architecture of the liver. When we select a particular method, it is necessary to consider the actual convenience for using the method, technical limitations of each method (maximum scale, detectable cell types, compatibility with fluorescent proteins such as GFP, etc.), and the resolution. Among them, resolution should be particularly important to visualize the actual tissue structures. In fact, the analysis of the biliary tree architecture with an unprecedented high-resolution analysis revealed the actual heterogeneous structure (Fig. 3) and the dynamic rearrangements of the bile ducts [6, 9]. In contrast, as shown in Fig. 3, it is hard to know the correct tissue structures and their changes by low-resolution analyses. Since only large ducts are observed in a low-resolution analysis, bile ducts are mostly depicted in textbooks as a single tube along the vein. Thus, high-resolution analyses do not only lead to new discoveries but are also crucial for getting the correct conclusions.

In this chapter, we will introduce macroscopic and microscopic 3D analysis methods, which should preferably be combined together to cover most of our purposes for analyzing the tissue

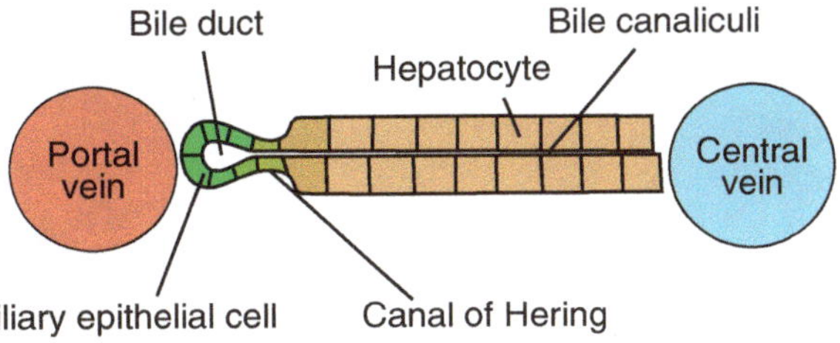

Fig. 2 The continuous epithelial tissue structure of the liver. The epithelial tissue of the liver consists of hepatocytes and biliary epithelial cells, which are lined up as a continuous structure, connected at the canal of Hering. Bile juice is produced by hepatocytes and drained into bile ducts through bile canaliculi. Hepatocytes near the bile ducts are known to have different characteristics from those located away from the bile duct

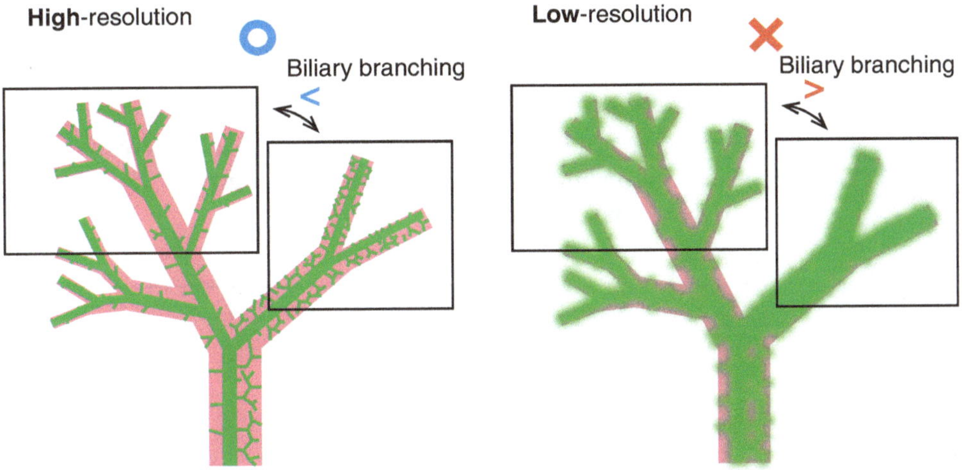

Fig. 3 Structure of the biliary tree (green) with the portal vein (pink). Bile ducts have heterogeneous structure consisting of thick tubes running along with the portal vein and fine tubes branching and wrapping around the portal vein. High-resolution analysis clearly shows the heterogeneous network (left panel), whereas the fine tubular network is hardly visualized with low-resolution analysis (right panel). Therefore, they possibly lead to different conclusions

structure during regeneration. Macroscopic analysis visualizes the whole vasculature and the biliary tree structure in the entire mouse liver (Fig. 4). Briefly, ink is injected into the biliary tree retrogradely through the extrahepatic biliary tree using a glass capillary. Then the harvested liver is made optically transparent using an organic reagent. Microscopic analysis visualizes all kinds of structures in high resolution as long as antibodies are available. Objects in thick

Fig. 4 Macroscopic observation of the liver tissue structure by ink injection. Bile ducts (black) are running along and wrapping around the portal vein. Portal veins (PV) and central veins (CV) can be seen with the red blood in them

sections are labeled by immunostaining and imaged by optical sectioning and reconstituting with a confocal microscope (Fig. 5), which gives much higher resolution along the *Z*-axis as compared to stacking serial sections.

2 Materials

2.1 Macroscopic Analysis

1. Mice.
2. Hematocrit tubes for making capillaries (Fig. 6): Microhematocrit capillary tubes (22-362-566; Fisher Scientific) work well, though any other glass tube might work.
3. Syringe: Syringes with smaller volumes (typically around 1 mL) are better for controlling the air pressure.
4. Tube to connect the glass capillary with a syringe: Cut a butterfly needle to get a tube with a socket to connect with a syringe (Fig. 6).
5. Ink: Black waterproof fountain pen ink, SPC-200 (Platinum Japan), is suitable for the visualization (*see* **Note 1**).
6. PBS.
7. Ethanol.
8. BABB: Mix 500 mL of benzyl benzoate and 250 mL of benzyl alcohol (*see* **Note 2**).
9. 20% sucrose in PBS.
10. O.C.T. compound.

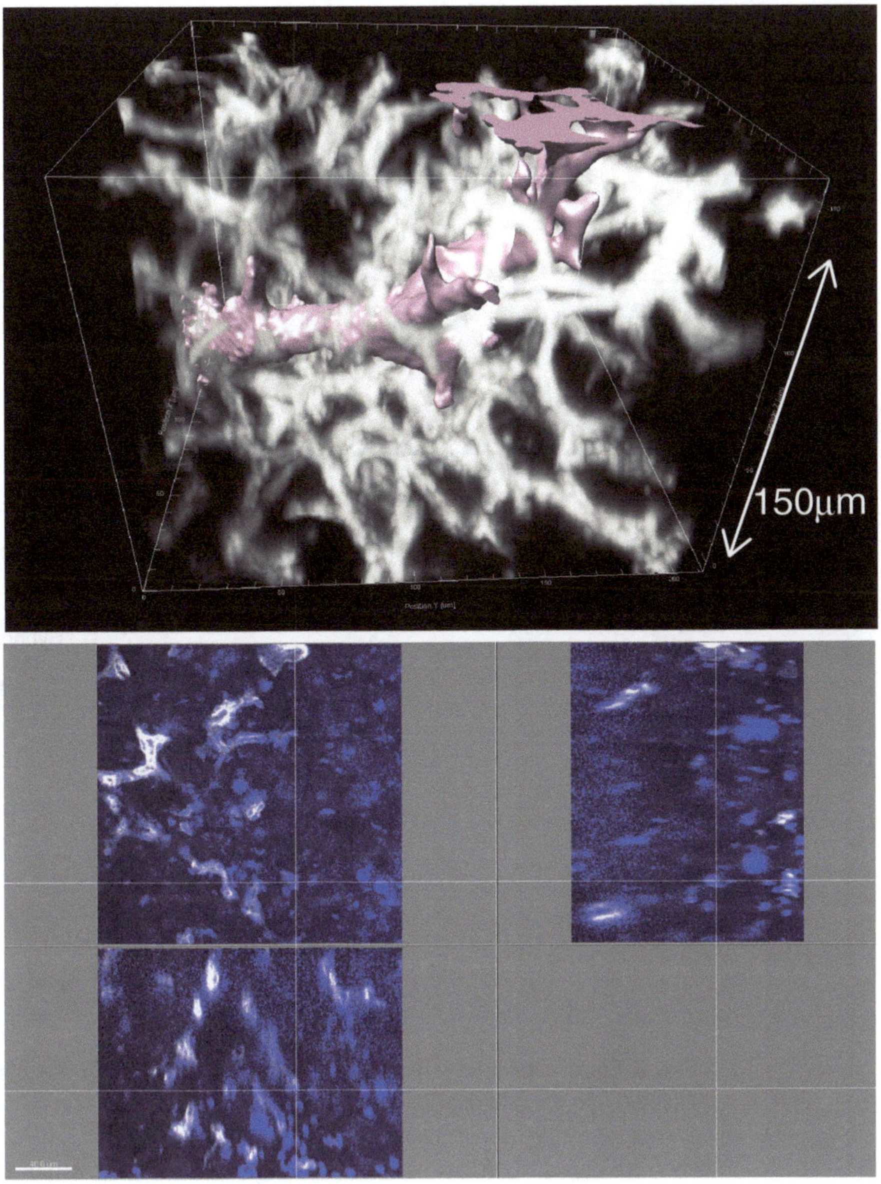

Fig. 5 Microscopic 3D reconstitution of the liver tissue structure. This image shows the biliary epithelial cells (white) in the CDE diet model liver. Veins (pink) can be reconstituted by inverting the autofluorescence of the tissue. Since this method is based on immunostaining, it can show subcellular structures of any types of liver cells as long as antibodies are available

2.2 Microscopic Analysis

1. Frozen liver tissue (*see* **Note 3**).
2. Antibody reaction solution: Dissolve antibodies in PBS containing 5% BSA, 0.1% Triton X-100, and 0.01% sodium azide. This reagent can be reused until the antibodies are consumed.

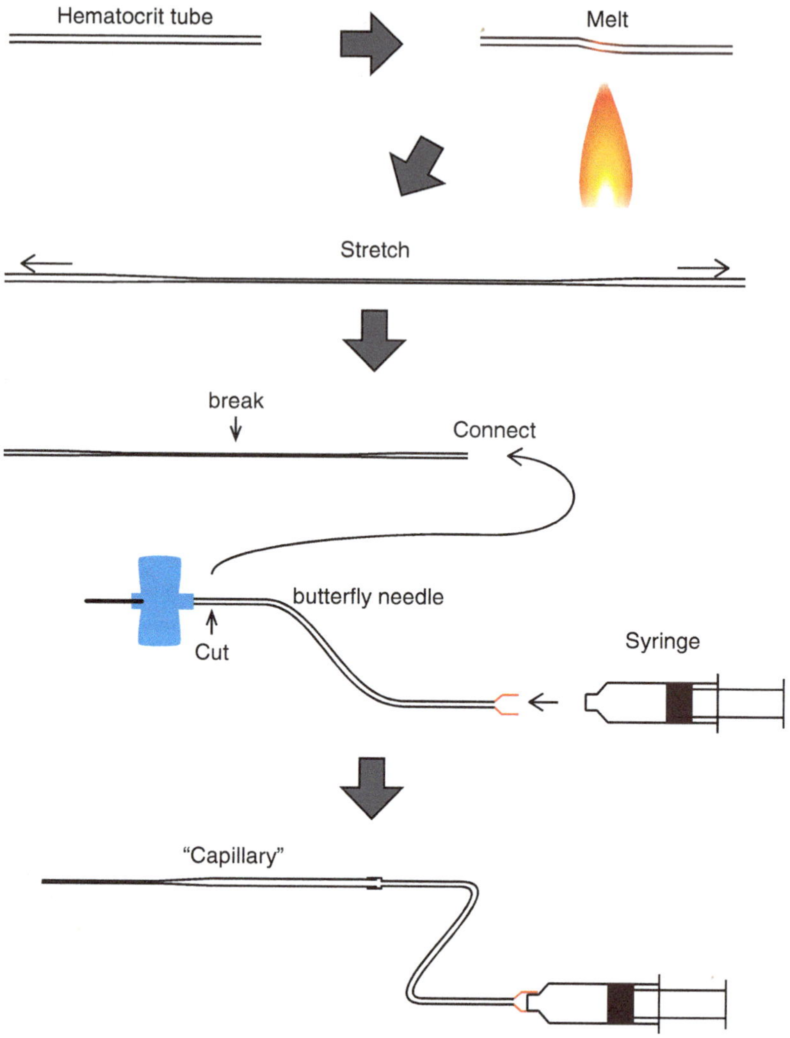

Fig. 6 Glass capillary for injection. Ink can be loaded into the capillary and injected into the bile duct by manipulating the syringe

The concentration of the antibodies should be similar to that for a regular immunostaining or half of that. The total volume needed for one thick section is usually 500–1000 μL.

3. Washing buffer: PBS containing 0.1% Triton X-100.
4. Fixation buffer: 4% PFA in PBS (pH 7.4) or 2% PFA 15% picric acid in PBS (pH 7.4).
5. Denaturing buffer: PBS containing 5% β-mercaptoethanol (ME) and 0.1% Triton X-100.
6. Counter staining solution: PBS containing 0.1% Triton X-100 with dyes such as DAPI or Hoechst.

7. Clearing/mounting reagent: Choose one from 2,2-′-thiodiethanol (TDE), saturated sucrose solution in PBS, or SeeDB [10].
8. Confocal microscope.

3 Methods

3.1 Macroscopic Analysis

1. Prepare capillaries (Fig. 6) with different thicknesses by melting the hematocrit tube in the middle with a gas burner or a candle flame and pulling the both ends apart to stretch the tube (*see* **Note 4**). Once getting the thinner part in the middle of the tube, break the part to make a capillary for insertion into the extrahepatic bile duct.
2. Euthanize the mouse, and open the abdominal cavity. Expose the liver, the portal vein, and the extrahepatic bile duct.
3. Identify the junction between the extrahepatic bile duct tube and the intestine (Fig. 7a). Select a capillary with a proper thickness that fits the bile duct tube (*see* **Note 5**).
4. Connect the glass capillary to a tube from a butterfly needle connected to a syringe (Fig. 6). Load the capillary with the ink (*see* **Note 6**).
5. Cut the junction between the bile duct and the intestine to make a hole for insertion of the capillary (Fig. 7b).
6. Pull the bile duct tube with sharp tweezers without blocking the hole, and insert the capillary loaded with the ink (Fig. 7c) (*see* **Note 7**).
7. Inject the ink very slowly while carefully observing the surface of the liver. The gallbladder will turn black first, when the capillary is successfully inserted into the bile duct. Stop the injection, when tiny black dots appear on the surface of the liver (*see* **Note 8**).
8. Harvest the liver and immerse into PBS in a petri dish. Place the liver between cotton pads in the PBS to fix the liver posture, if desired.
9. Place the petri dish on a shaker and gradually add ethanol to the PBS in a petri dish. Increase the ethanol concentration until finally replacing the whole PBS with ethanol (*see* **Note 9**). For instance, PBS for 1 min, 15% EtOH in PBS for 10 min, 30% EtOH in PBS for 10 min, 40% EtOH in PBS for 20 min, 60% EtOH in PBS for 30 min, 80% EtOH in PBS for 30 min, and 100% EtOH for 30 min three times.
10. After complete dehydration, immerse the liver in BABB. Refresh the BABB solution until the liver becomes completely transparent (*see* **Notes 2** and **10**). With a whole liver in 50 mL

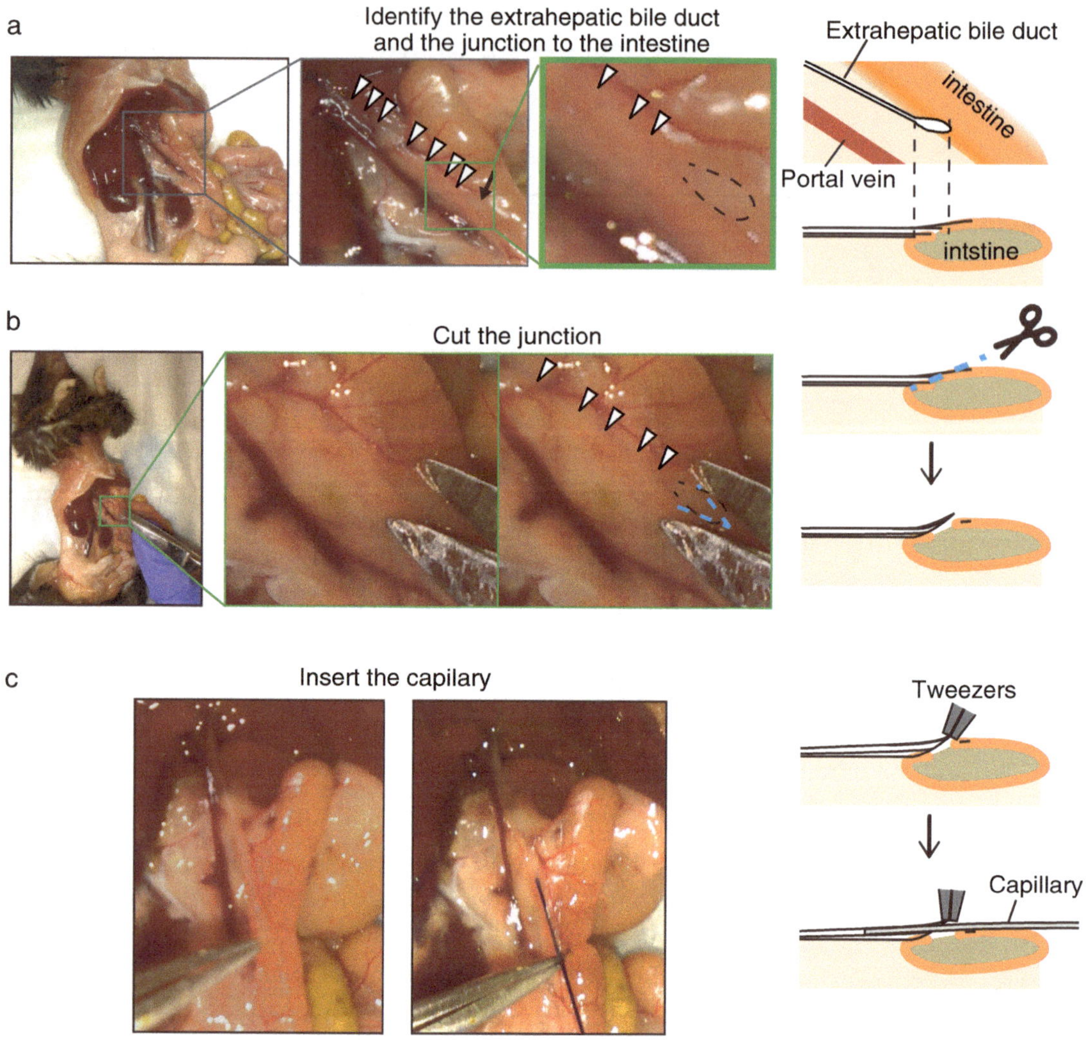

Fig. 7 Procedures of the ink injection into the biliary tree. (**a**) Extrahepatic bile duct (arrowheads) is running from the liver to the intestine on the side of the portal vein. The bile duct has a bulge (arrow and dashed line) at the junction with the intestine. (**b**) Cutting the bulge at the junction. (**c**) Insertion of the capillary into the bile duct tube

tube, the treatment in 20 mL BABB for 1 h is repeated three times.

11. Observe and take pictures of the liver (*see* **Note 11**).
12. The liver after **step 11** can be immersed back in ethanol and then PBS, 20% sucrose in PBS and be embedded in O.C.T. compound and frozen for sectioning and immunostaining.
13. Successful ink injection can be evaluated by the coexistence of biliary markers and ink on the liver sections.

3.2 Microscopic Analysis

1. Make sections of the liver in around 200 μm thickness by cryostat (*see* **Note 12**). Directly drop the frozen section in the cold PBS in 50 mL tube, and unroll the section by gently inverting the tube. If the tissue was embedded in O.C.T. compound, the tissue section is released from the compound during this step.
2. Substitute the PBS with 25 mL of cold washing buffer. Rock the tube for 1 h at 4 °C.
3. Discard the liquid, leaving 1 mL. With this 1 mL buffer, transfer the liver section to 5 mL tube or proper sized/shaped container. Discard all the liquid and add the fixation buffer and incubate for overnight at 4 °C, if the tissue is not prefixed. For prefixed tissues, add the denaturing buffer for antigen retrieval/permeabilization, and incubate for 1 h at 37 °C.
4. Transfer the section to 50 mL tube with the liquid, discard the liquid, and wash the section with 25 mL of cold washing buffer for more than 1 h three times and final wash for overnight, rocking.
5. Discard the liquid, leaving 1 mL. With this 1 mL buffer, transfer the liver section to 5 mL tube or proper sized/shaped container. Discard all the liquid and add the antibody reaction solution.
6. Incubate for more than 2 days at 4 °C, rocking the container (*see* **Note 13**). Occasional manual rocking and flipping the section gives better results.
7. Transfer the antibody reaction solution to another tube for reusing next time. Add 1 mL of cold washing buffer to the section and transfer them to 50 mL tube. Wash the section three times in 25 mL of cold washing buffer with rocking for more than 1 h.
8. Repeat **steps 5–7** for the reaction with secondary antibodies. Then wash the section with cold washing buffer overnight.
9. Discard the liquid, leaving 1 mL. With this 1 mL buffer, transfer the liver section to 5 mL tube or proper sized/shaped container. Discard all the liquid and add the counter-staining solution.
10. Wash the section with cold washing buffer half an hour three times.
11. Discard the liquid, leaving 1 mL. With this 1 mL buffer, transfer the liver section on to a glass slide. Remove the liquid with pipette, and add 50 μL of the clearing/mounting reagent. Replace the liquid with the clearing/mounting reagent until the section becomes evenly clear.

12. Cover the section with a glass cover slip (*see* **Note 14**), and acquire images with a confocal microscope.

4 Notes

1. The main component of this ink is "carbon black," fine particles of amorphous carbon dissolved in a detergent. When injected into the biliary tree, the carbon particles get stuck inside the biliary tube, while the solvent in the ink leaks out to the portal veins and the bile canaliculi. This makes the biliary tube filled with carbon particles. Thus, any other ink with the similar components may work, but the particle size of the carbon black, which differs according to the manufacturing, is crucial for getting the optimal result.
2. BABB melts plastics except polypropylene. Therefore glass or polypropylene should be used to contain BABB.
3. Liver tissues are fixed or non-fixed before frozen depending on the antibody used.
4. Once the glass melts and becomes very unsolid on the fire, displace the tube away from the flame to let it slightly solidify for around 0.2–1 s. When the melted part becomes "resilient" or "viscous," pull apart both ends of the tube. The thickness and length of the capillary are determined by the solidity of the melted part on the moment it is stretched. Longer capillary part with the similar thickness as the extrahepatic bile duct is desirable. Since the thickness of the extrahepatic bile duct differs with each mouse, it is important to prepare capillaries with different thicknesses.
5. It is important to use a capillary with a proper thickness that fits the thickness of the biliary tube. If the capillary is too thin and/or sharp, it easily penetrate through tissues so that it would get into wrong places or break the wall of the extrahepatic bile duct.
6. Ink should be carefully loaded on to the bottom half (or 80%) of the capillary/tube without ever passing the half (or 80%) point. Otherwise the capillary/tube will be coated with the dark ink, and the surface of the loaded ink becomes difficult to recognize. This surface is an important indicator of how fast the ink is moving while injecting it into the biliary tree. To prevent the overload, level the air pressure inside the syringe to zero or slightly positive before taking the capillary out of the ink. This prevents the suction of the air into the capillary, which happens much quicker than the suction of the ink.
7. When inserting the capillary into the bile duct tube, the air pressure inside the syringe should be zero or slightly positive to

avoid loading of air or any tissue fragments on the capillary tip. Poking tissues randomly with the capillary during the insertion into the bile duct may insert tissue fragments inside the capillary. This can cause clogs inside the intrahepatic biliary tree during ink injection, preventing uniform ink loading throughout the entire liver.

8. While injecting the ink, adjusting the posture of the liver may help the well-distributed injection over the entire liver. Excessive ink load leads to leakage of the ink (carbon particle) to the bile canaliculi and especially portal veins. The amount of the ink needed for one adult mouse is usually around 15–20 μL, but it depends on every individual, and the injection amount should be determined based on observation of the liver surface during injection. Tiny black dots appearing on the liver surface are a sign to stop injecting.
9. Rapid increase of the ethanol concentration will make strongly dehydrated parts which block penetration of the liquid into other parts, leading to an insufficient dehydration. It is important to make sure the color of the whole liver is gradually turning to white evenly as dehydration proceeds. Sufficient dehydration is curial to clear the tissue by BABB.
10. Used BABB contaminated with ethanol can be reused for initial steps of clearing in **step 10**.
11. Blood remained in the vasculature visualizes the portal veins and the central veins by the transparent red color. The portal veins and the central veins can easily be distinguished by the extrahepatic bile duct entering the liver along with the portal vein. Vasculature can also be stained with perfusion of any ink before the ink injection into the biliary tree. However, strong stains on the vasculature make the observation difficult, due to the dense existence of the vasculature in the whole liver.
12. Higher cutting temperature makes it easier to get thick sections. The temperature depends on the cryosectioning machine used, but it is usually optimal around −15 °C.
13. Incubation with primary antibodies can be performed longer (3 days to 1 week) for better results, while incubation with secondary antibodies may be optimal with just 2 days.
14. Insert something (glass cover slips work fine) between the cover slip and the glass slide near the edge of the cover slip to avoid pressing the thick section by the cover slip.

Acknowledgments

The experiments to establish these methods were supported in part by Japan Society for the Promotion of Science KAKENHI grants

22118006 (to Dr. Atsushi Miyajima), 24112507 (to Dr. Tohru Itoh), 24590342 (to T.I.), and 26253023 (to A.M.) as well as by CREST from the Japan Science and Technology Agency (to A.M.). K. K. was a research fellow at the Japan Society for the Promotion of Science.

References

1. Miyajima A, Tanaka M, Itoh T (2014) Stem/progenitor cells in liver development, homeostasis, regeneration, and reprogramming. Cell Stem Cell 14(5):561–574. https://doi.org/10.1016/j.stem.2014.04.010
2. Font-Burgada J, Shalapour S, Ramaswamy S, Hsueh B, Rossell D, Umemura A, Taniguchi K, Nakagawa H, Valasek Mark A, Ye L, Kopp Janel L, Sander M, Carter H, Deisseroth K, Verma Inder M, Karin M (2015) Hybrid periportal hepatocytes regenerate the injured liver without giving rise to cancer. Cell 162(4):766–779. https://doi.org/10.1016/j.cell.2015.07.026
3. Tanimizu N, Ichinohe N, Yamamoto M, Akiyama H, Nishikawa Y, Mitaka T (2017) Progressive induction of hepatocyte progenitor cells in chronically injured liver. Sci Rep 7:39990. https://doi.org/10.1038/srep39990
4. Wang B, Zhao L, Fish M, Logan CY, Nusse R (2015) Self-renewing diploid Axin2(+) cells fuel homeostatic renewal of the liver. Nature 524(7564):180–185. https://doi.org/10.1038/nature14863
5. Roskams TA, Theise ND, Balabaud C, Bhagat G, Bhathal PS, Bioulac-Sage P, Brunt EM, Crawford JM, Crosby HA, Desmet V, Finegold MJ, Geller SA, Gouw AS, Hytiroglou P, Knisely AS, Kojiro M, Lefkowitch JH, Nakanuma Y, Olynyk JK, Park YN, Portmann B, Saxena R, Scheuer PJ, Strain AJ, Thung SN, Wanless IR, West AB (2004) Nomenclature of the finer branches of the biliary tree: canals, ductules, and ductular reactions in human livers. Hepatology (Baltimore, MD) 39(6):1739–1745. https://doi.org/10.1002/hep.20130
6. Kaneko K, Kamimoto K, Miyajima A, Itoh T (2015) Adaptive remodeling of the biliary architecture underlies liver homeostasis. Hepatology (Baltimore, MD) 61(6):2056–2066. https://doi.org/10.1002/hep.27685
7. Kamimoto K, Kaneko K, Kok CYY, Okada H, Miyajima A, Itoh T (2016) Heterogeneity and stochastic growth regulation of biliary epithelial cells dictate dynamic epithelial tissue remodeling. eLife 5:e15034
8. Nagahama Y, Sone M, Chen X, Okada Y, Yamamoto M, Xin B, Matsuo Y, Komatsu M, Suzuki A, Enomoto K, Nishikawa Y. Contributions of hepatocytes and bile ductular cells in ductular reactions and remodeling of the biliary system after chronic liver injury. Am J Pathol 184(11):3001–3012. https://doi.org/10.1016/j.ajpath.2014.07.005
9. Tanimizu N, Kaneko K, Itoh T, Ichinohe N, Ishii M, Mizuguchi T, Hirata K, Miyajima A, Mitaka T (2016) Intrahepatic bile ducts are developed through formation of homogeneous continuous luminal network and its dynamic rearrangement in mice. Hepatology (Baltimore, MD) 64(1):175–188. https://doi.org/10.1002/hep.28521
10. Ke M-T, Fujimoto S, Imai T (2013) SeeDB: a simple and morphology-preserving optical clearing agent for neuronal circuit reconstruction. Nat Neurosci 16:1154. https://doi.org/10.1038/nn.3447

Part V

Liver Injury Models

Chapter 18

Canine Liver Fibrosis Model to Assess the Functions of Infused Autologous Bone Marrow-Derived Cells

Taro Takami, Kenji Tani, Yasuho Taura, and Isao Sakaida

Abstract

Liver cirrhosis is the end stage of chronic liver disease, and the only radical treatment for decompensated liver cirrhosis is still liver transplantation. Development of effective regenerative therapy for liver cirrhosis is an urgent task. Before human clinical trials can be considered, the safety and efficacy of any planned protocol must be confirmed in medium-to-large animals. Therefore, we have developed a novel canine liver fibrosis model for proof of concept (POC) studies.

Key words Liver cirrhosis, Canine model, Mesenchymal stem cells, Carbon tetrachloride

1 Introduction

Liver cirrhosis (LC) is the end stage of chronic liver disease caused by viral hepatitis, excess consumption of alcohol, obesity, and so on. Although the liver is an organ with a high regenerative capacity, chronic inflammation results in the development of cirrhosis, and dense fibrosis ultimately halts hepatocyte proliferation. Currently, the only radical treatment available for decompensated LC is liver transplantation. However, liver transplantation has its own issues, including donor shortage, immunorejection, and medical cost. To overcome the issues related to liver transplantation, we aimed to develop a liver regeneration therapy for decompensated LC patients using autologous bone marrow cells (BMCs) [1]. In a murine LC model created by repeated carbon tetrachloride (CCl_4) treatment [2], we previously confirmed that syngeneic non-cultured whole BMCs that were infused through the tail vein had repopulated the damaged liver and produced matrix metalloproteinases (MMPs), ultimately reducing liver fibrosis [3]. Based on these evidences, we conducted a clinical study and reported the clinical efficacy and safety of "autologous bone marrow cell infusion therapy (ABM*i* therapy)" as liver regeneration therapy for decompensated LC

Naoki Tanimizu (ed.), *Hepatic Stem Cells: Methods and Protocols*, Methods in Molecular Biology, vol. 1905, https://doi.org/10.1007/978-1-4939-8961-4_18, © Springer Science+Business Media, LLC, part of Springer Nature 2019

[4–6]. However, the therapy involves bone marrow (BM) aspiration under general anesthesia. Therefore, we aimed to develop a less invasive therapy that uses cultured autologous BM-derived mesenchymal stem cells (BMSCs) aspirated under local anesthesia. We previously demonstrated that the peripheral infusion of cultured BMSCs reduced liver fibrosis in an immunodeficient cirrhotic murine model [7] and thioacetamide (TAA)-induced cirrhotic murine model [8]. However, before human clinical trials can be considered, the safety and efficacy of the protocol must be confirmed in medium-to-large animals. Therefore, we aimed to develop a canine liver fibrosis model by CCl_4 treatment. It was difficult to make a canine take CCl_4 orally, because of a keen nose. We have finally established a novel canine liver fibrosis model by repeated direct CCl_4 injections into the stomach to demonstrate the safety and efficacy of infusion of autologous BMCs for the treatment of LC [9]. Here, we describe the method of creating this canine liver fibrosis model.

2 Materials

2.1 Gastric Catheter Placement

1. Beagles (1–2 years old): These canines were housed in our animal facility and treated in accordance with the university's animal care guidelines. This study was approved by our Institutional Ethics Committee (approval number; 21-033).
2. Infusion port: A P-U Celsite port (Toray Medical Co. Ltd., Tokyo) consisting of a catheter coated with a heparinized hydrophilic material and a Celsite port.
3. 18 G Teflon IV catheter (6-French P-U catheter).
4. Anesthesia machine: For delivering positive pressure ventilation.
5. Propofol 1% injection: The loading dose of propofol is about 7 mg/kg body weight (BW).
6. Lepetan: Intravenous Lepetan (buprenorphine) is administered at a dose of 10 μg/kg BW to provide effective pain relief.
7. Isoflurane: Anesthesia is maintained with isoflurane in oxygen. The end-tidal isoflurane concentration is monitored and maintained between 1.4 and 2.8%.
8. Endobronchial tube for general anesthesia.
9. Chlorhexidine acetate.
10. 2% chlorhexidine gluconate in 70% isopropyl alcohol.
11. #21 scalpel.
12. Right-angled forceps.
13. Babcock forceps.

14. Suture material (3-0 PDS).
15. Povidone-iodine.
16. Upper gastrointestinal endoscope (Olympus Co., Tokyo).

2.2 Aspiration of Bone Marrow Fluid, Culture, and Cell Infusion of Bone Marrow Derived-Mesenchymal Stem Cells (BMSCs)

1. Medetomidine.
2. Atipamezole hydrochloride.
3. 16 G biopsy needle.
4. 5- and 20-mL syringe.
5. Heparin solution.
6. Saline.
7. Culture medium: 10% fetal bovine serum and 100 μg/mL gentamicin in Dulbecco's Modified Eagle Medium (DMEM).
8. Cell detachment solution: TrypLE Select (Thermo Fisher Scientific Inc.).
9. Phosphate-buffered saline (PBS).
10. Culture flask: T-75 flasks.

2.3 Induction of Liver Fibrosis

1. Carbon tetrachloride (CCl_4).
2. Corn oil.
3. Winged Surecan port needle, 22 G × 25 mm.

2.4 Liver Biopsy

1. Medetomidine.
2. Atipamezole hydrochloride.
3. Isoflurane: Anesthesia is maintained with isoflurane in oxygen. The end-tidal isoflurane concentration is monitored and maintained between 1.4 and 2.8%.
4. 0.5% lidocaine for local anesthesia.
5. Povidone-iodine.
6. 16 G biopsy needle (Aragon Medical Devices, Plano, TX).
7. Ultrasound system: This consists of an ultrasound machine (Aplio MX, Canon Medical Systems Co., Tokyo), probe (PVT-745BTV convex electronic scanning probe, Canon Medical Systems Co., Tokyo), and needle biopsy adaptor (UAGV-029A, Canon Medical Systems Co., Tokyo).

2.5 Indocyanine Green (ICG) Test

1. Indocyanine green.
2. ICG meter A-30 (Fuchu Giken, Inc., Tokyo).
3. 10 unit/mL heparin: 1000 units of heparin are mixed with 100 mL of physiological saline.

2.6 Sirius Red Staining

1. Sirius red solution (1.3% aqueous picric acid/0.1% direct red 80).

2. 0.5% acetic acid.
3. 100% ethanol.
4. Xylene.

3 Methods

3.1 Gastric Catheter Placement

1. Under general anesthesia, the canine is restrained in the right lateral decubitus position and shaved over a wide area from the posterior margin of the scapula to the anterior margin of the femur and from the linea alba (midline) to around 2 cm left of the vertebral body.
2. Wearing sterile gloves, the operator picks up a chlorhexidine acetate scrub in a piece of gauze and uses it to scrub the operating field, foaming it thoroughly, and disinfects the area with cotton soaked in 2% chlorhexidine gluconate in 70% isopropyl alcohol. The operating field is then sprayed with povidone-iodine and allowed to dry naturally, after which it is draped.
3. An approximately 3-cm-long skin incision is made with a #21 scalpel at a point 2 cm caudal to the last rib, blunt dissection of the subcutaneous tissue is performed using right-angled forceps, and the abdominal external oblique muscle is grasped with Babcock forceps and exposed. Right-angled forceps are used to divide the muscles bluntly, and the abdominal internal oblique muscle is similarly divided to approach the abdominal cavity.
4. An assistant operates the upper gastrointestinal endoscope and inflates the stomach, assisting the operator by making it easier to grasp the serosal side of the stomach with Babcock forceps. Part of the stomach is exposed extracorporeally and the stomach is fixed to the abdominal wall by placing single ligatures of suture material (3-0 PDS) at four places on the gastric wall; then, the abdominal external oblique muscle around the location is grasped with the Babcock forceps.
5. The catheter is inserted into the stomach using a sheath, guidewire, and dilator (Fig. 1a). Endoscopy is used to confirm that the catheter tip does not extend past the cardiac orifice into the esophagus (Fig. 1b).
6. An approximately 3-cm-long skin incision is made slightly to the left of the eighth to tenth thoracic vertebrae, and blunt dissection of the subcutaneous tissues is performed to create a bed for the port (Fig. 1c). An approximately 1-cm-long skin incision is made at a point exactly midway between the sites of port placement and catheter insertion, and right-angled forceps are passed through this incision to grasp the catheter and guide it under the skin to the port placement site.

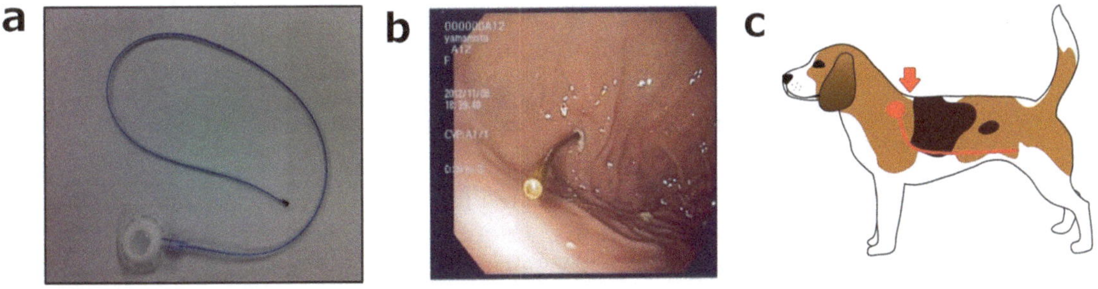

Fig. 1 Canine liver fibrosis model. (**a**) Infusion port: A P-U Celsite port. (**b**) The catheter tip in the stomach was confirmed using endoscopy. (**c**) The port was placed in the back of the neck. Arrow shows the place of the port

7. The length of the catheter is adjusted, and the catheter is connected to the port via a device that joins and anchors the catheter and port.
8. The port is placed under the skin and subcutaneous tissue and anchored by suturing it to the subcutaneous tissue with 3-0 PDS. The skin incisions are then sutured with single ligatures of 3-0 PDS.

3.2 Aspiration of Bone Marrow Fluid, Culture, and Cell Infusion of BMSCs

1. After intramuscular injection of 20 μg/kg medetomidine, the canine is restrained in the right lateral decubitus position, and an area approximately 4 cm in diameter in the center of the left shoulder is shaved and disinfected.
2. The greater tubercle of the humerus is palpated, and after the aspiration site has been decided, a tiny skin incision is made with an 18 G needle.
3. After the humerus has been firmly immobilized with the right hand, a 16 G biopsy needle is inserted into the greater tubercle. Once it has passed through the cortical bone, the inner needle is withdrawn, and the bone marrow is aspirated using a 5-mL syringe filled with 1 mL of heparin.
4. The 16 G biopsy needle is withdrawn and manual compression is applied for hemostasis, and then atipamezole hydrochloride is intramuscular injected finally.
5. To culture bone marrow derived-mesenchymal stem cells (BMSCs), the aspirated BM fluid is seeded into T-75 flasks and cultured in culture medium in a 5% CO_2 incubator at 37 °C.
6. After 2 days of incubation, nonadherent cells are removed during medium replacement. The culture medium is changed every 2 days for 2–3 weeks.
7. Expanded BMSCs are detached using cell detachment solution and then are collected. BMSCs are washed with PBS and resuspended in 20 mL saline. BMSCs are injected through the peripheral vein slowly (around 2–5 mL/min) using 20-mL syringe.

3.3 Induction of Liver Fibrosis

1. CCl_4 is diluted 1:1 in corn oil before its intragastric injection through the catheter.
2. A winged Surecan port needle is inserted into the subcutaneous port, and the mixture of carbon tetrachloride and corn oil is injected into the stomach (*see* **Note 1**).
3. Injections are repeated for 10 weeks using the implanted catheter (high-dose period, 1.0 mL/kg BW once a week and 0.5 mL/kg BW once a week for 6 weeks; low-dose period, 0.25 mL/kg BW twice a week for 4 weeks) to induce liver fibrosis [9]. After the infusion of BMSCs, low-dose CCl_4 is further administrated for 4 weeks.
4. The CCl_4 injections are made on the first and fourth day of each week.

3.4 Liver Biopsy

1. After intramuscular injection of 20 μg/kg medetomidine, the canines are placed in the left lateral decubitus position.
2. The skin over the biopsy site is shaved, followed by disinfection with povidone-iodine.
3. The local anesthetic (0.5% lidocaine) is injected into a small area of the skin and tissues over part of the liver, and ultrasonography-guided liver biopsies are carried out using a 16 G biopsy needle (Aragon Medical Devices, Plano, TX) (*see* **Note 2**).
4. After removing the biopsy needle, the puncture site is manually compressed, and then atipamezole hydrochloride is intramuscular injected finally.
5. Some samples are fixed in 4% paraformaldehyde overnight and used for the evaluation of liver fibrosis by Sirius red staining (Fig. 2a–c), while others are stored at −80 °C for microarray analysis and real-time quantitative polymerase chain reaction (RT-qPCR) testing.

3.5 Indocyanine Green (ICG) Test to Assess the Liver Functions

1. A 22 G indwelling venous cannula is inserted into the internal jugular vein and blood is collected. This venous route is kept open with physiological saline (*see* **Notes 3–5**).
2. ICG is prepared immediately prior to its administration.
3. Following the collection of blood for the zero-time sample, 0.1 mg ICG/kg is administered via the peripheral vein, and blood is collected 5, 10, 15, 20, and 30 min later from the maintained venous route (*see* **Note 3**).
4. Blood samples are centrifuged.
5. Serum is harvested from blood samples and analyzed for ICG content using an ICG meter (Fig. 2d) (*see* **Note 5**).

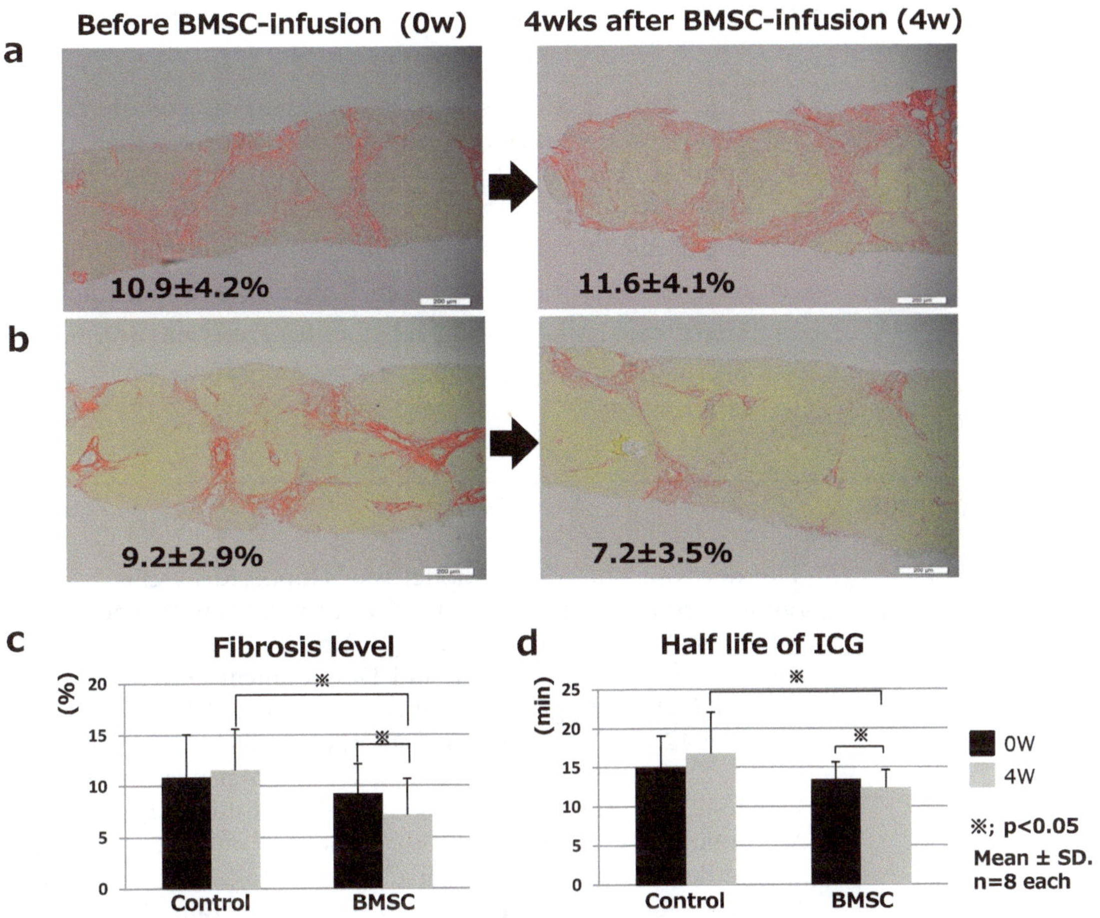

Fig. 2 Kinetics of liver fibrosis and half-life time of indocyanine green (ICG) testing after the infusion of cultured autologous bone marrow derived-mesenchymal stem cells (BMSCs) in the canine liver fibrosis model. (**a**, **c**) The amount of fibrosis, as seen with Sirius red staining, was increased in the control group (after continuous carbon tetrachloride treatment without the autologous BMSC infusion). (**b**, **c**) A decreased level of fibrosis was seen at 4 weeks after the autologous BMSC infusion, even though carbon tetrachloride treatment was also continued. (**d**) The half-life of ICG was significantly shorter at 4 weeks in the BMSC-infused group compared with 0 weeks in the same group and 4 weeks in the control group. Every value and bar shown indicates mean ± SD. This figure was reproduced with permission from ref. [9]

3.6 Sirius Red Staining

1. Dewax and hydrate paraffin sections.
2. Stain liver fibrosis with Sirius red solution for 15–20 min.
3. Wash with acetic acid for 1 min.
4. Dehydrate with 100% ethanol for 5 min.
5. Clear in xylene for 10 min and then mount in a resinous medium.

4 Notes

1. When CCl_4 is administered, the port and catheter are flushed with water to confirm the absence of resistance before administration. The catheter is finally flushed out with water to ensure that no food residue becomes stuck in the catheter.
2. The needle insertion angle for liver biopsy is usually 44°.
3. Around 5 mL of blood is sampled after ICG administration.
4. Animals are fasted from 12 h before ICG tests and liver biopsy.
5. The amount of serum used for the ICG meter is 500 μL each time.

Acknowledgments

The development of this model was supported by the Japanese highway program for the realization of regenerative medicine of the Japan Science and Technology Agency (JST) and the Japan Agency for Medical Research and Development (AMED) and by JSPS KAKENHI grant numbers JP17H04162 to I.S. and JP17K09428 to T.T. We also thank the members involved in this model, Luiz Fernando Quintanilha, Takashi Matsuda, Bruno Diaz Paredes, Yuki Aibe, Ryo Sasaki, Tatsuro Nishimura, Toshihiko Matsumoto, Tsuyoshi Ishikawa, Naoki Yamamoto, Yasuhisa Oishi, Ikuyo Yoshida, Koushiro Ozono, Takafumi Kanematsu, Katsuhiro Kesamaru, Hironori Morisaki, Miku Yamashita, Yutaro Kitaura, Ko Nakasumi, Kazuaki Igari, Shotaro Eto, Kyosuke Hidari, Masahiro Imagawa, Tsunehiro Isozaki, Kentaro Itoh, Haruhi Sakai, Mai Sakai, Hiroya Inoue, Fumiya Yamane, Masato Hiyama, and Saori Shinzato.

References

1. Takami T, Terai S, Sakaida I (2012) Stem cell therapy in chronic liver disease. Curr Opin Gastroenterol 28(3):203–208. https://doi.org/10.1097/MOG.0b013e3283521d6a
2. Terai S, Sakaida I, Yamamoto N, Omori K, Watanabe T, Ohata S, Katada T, Miyamoto K, Shinoda K, Nishina H, Okita K (2003) An in vivo model for monitoring transdifferentiation of bone marrow cells into functional hepatocytes. J Biochem 134(4):551–558
3. Sakaida I, Terai S, Yamamoto N, Aoyama K, Ishikawa T, Nishina H, Okita K (2004) Transplantation of bone marrow cells reduces CCl4-induced liver fibrosis in mice. Hepatology 40 (6):1304–1311. https://doi.org/10.1002/hep.20452
4. Terai S, Ishikawa T, Omori K, Aoyama K, Marumoto Y, Urata Y, Yokoyama Y, Uchida K, Yamasaki T, Fujii Y, Okita K, Sakaida I (2006) Improved liver function in patients with liver cirrhosis after autologous bone marrow cell infusion therapy. Stem Cells 24(10):2292–2298. https://doi.org/10.1634/stemcells.2005-0542. pii: 2005-0542
5. Kim JK, Park YN, Kim JS, Park MS, Paik YH, Seok JY, Chung YE, Kim HO, Kim KS, Ahn SH, Kim do Y, Kim MJ, Lee KS, Chon CY, Kim SJ, Terai S, Sakaida I, Han KH (2010) Autologous bone marrow infusion activates the progenitor cell compartment in patients with advanced liver cirrhosis. Cell Transplant 19(10):1237–1246. https://doi.org/10.3727/096368910X506863

6. Kim JK, Kim SJ, Kim Y, Chung YE, Park YN, Kim HO, Kim JS, Park MS, Sakaida I, Kim DY, Lee JI, Ahn SH, Lee KS, Han KH (2017) Long-term follow-up of patients after autologous bone marrow cell infusion for decompensated liver cirrhosis. Cell Transplant 26(6):1059–1066. https://doi.org/10.3727/096368917X694778
7. Tanimoto H, Terai S, Taro T, Murata Y, Fujisawa K, Yamamoto N, Sakaida I (2013) Improvement of liver fibrosis by infusion of cultured cells derived from human bone marrow. Cell Tissue Res 354(3):717–728. https://doi.org/10.1007/s00441-013-1727-2
8. Quintanilha LF, Takami T, Hirose Y, Fujisawa K, Murata Y, Yamamoto N, Goldenberg RC, Terai S, Sakaida I (2014) Canine mesenchymal stem cells show antioxidant properties against thioacetamide-induced liver injury in vitro and in vivo. Hepatol Res 44(10):E206–E217. https://doi.org/10.1111/hepr.12204
9. Matsuda T, Takami T, Sasaki R, Nishimura T, Aibe Y, Paredes BD, Quintanilha LF, Matsumoto T, Ishikawa T, Yamamoto N, Tani K, Terai S, Taura Y, Sakaida I (2017) A canine liver fibrosis model to develop a therapy for liver cirrhosis using cultured bone marrow-derived cells. Hepatol Commun 1(7):691–703. https://doi.org/10.1002/hep4.1071

Chapter 19

A Rodent Model for Cell Transplantation of Hepatic Progenitor Cells

Sei Kakinuma and Akihide Kamiya

Abstract

Hepatic progenitor cells are defined as cells exhibiting potency for active proliferation and capacity for bipotential differentiation into hepatocytes and cholangiocytes. To prove the capacity of target cells for terminal differentiation and reconstitution of organs, cell transplantation models have been widely used in previous studies, including those involving the liver. Here we describe a protocol for transplantation of hepatic progenitor cells using retrorsine pretreatment and partial hepatectomy. This transplantation assay reveals the potential for reconstitution of hepatocytes in recipient livers by primary hepatic progenitor cells. Donor cells are detected as a colony composed of 5–10 mature hepatocytes.

Key words Hepatoblasts, Hepatocytes, Intrasplenic injection, Partial hepatectomy, Retrorsine, Transplantation, Differentiation

1 Introduction

Stem cells are defined as cells exhibiting both the capacity for self-renewal and the ability to undergo multilineage differentiation [1]. Somatic stem cells in the liver (i.e., hepatic stem cells) are defined as cells capable of self-renewal and terminal differentiation into hepatocytes and cholangiocytes. Liver lobules are constituted of a variety of cells, including hepatocytes, cholangiocytes, hepatic stellate cells (Ito cells), Kupffer cells, sinusoidal endothelial cells, and pit cells, among others. Both hepatocytes and cholangiocytes are derived from hepatoblasts in the fetal liver [2]. Hepatic progenitor cells are also capable of bipotential differentiation and active proliferation but lack the capacity for self-renewal [3]. Hepatic stem/progenitor cells can be enriched in mouse mid-gestational fetal hepatic cell fractions based on several cell surface markers, including Dlk [4], E-cadherin [5], Liv2 [6], and CD13 [7]. Rat Dlk^+ or $Thy\text{-}1^+$ cells isolated from mid-gestational fetal liver exhibit characteristics expected for hepatic stem/progenitor cells [8]. In addition to enrichment in fetal liver, hepatic stem/progenitor cells

Naoki Tanimizu (ed.), *Hepatic Stem Cells: Methods and Protocols*, Methods in Molecular Biology, vol. 1905, https://doi.org/10.1007/978-1-4939-8961-4_19,

can also be enriched in mouse hepatic cells based on the expression of several molecules, including $CD13^+/CD49f^+/CD133^+$ cells in neonatal liver [9], $MIC1\text{-}1C3^+/CD133^+$ cells in adult healthy liver [10], and $Lgr5^+$ cells in adult damaged liver [11].

To verify the capacity of target cells for terminal differentiation and organ reconstitution, cell transplantation models involving various organ systems including blood (bone marrow), neuron, retina, intestine, and liver have been utilized. The objective of transplantation experiments with hepatic progenitor cells is to demonstrate the capacity for hepatocytic or cholangiocytic reconstitution by donor cells. In this section, we focus on the reconstitution of hepatocytes by transplantation of hepatic progenitor cells in rodent models. It is essential to inhibit proliferation of intrinsic mature hepatocytes for successful reconstitution by donor cells, because little or no engrafted donor cells are detected in a healthy liver or regenerating liver after simple partial hepatectomy in rodents. Previous studies have demonstrated that several strains of genetically modified mice can be used as recipients for hepatic cell transplantation, including transgenic (Tg) mice carrying an albumin promoter linked to mouse urokinase-type plasminogen activator [12] and fumarylacetoacetate hydrolase-deficient mice [13]. Intrinsic hepatocytes are continuously damaged in those mice, whereas transplanted donor cells are not damaged and can actively proliferate in the recipient liver. Alternatively, pretreatment of retrorsine also attenuates hepatocyte proliferation after partial hepatectomy, which allows donor cells to proliferate in the recipient liver.

To detect the donor cells in recipient livers, donor cells are labeled with marker proteins, such as green fluorescent protein (GFP), other fluorescent proteins, or β-galactosidase. The ratio of donor cells to recipient cells is assessed by histological analysis or secreted marker proteins in recipient serum. However, it is possible that overexpression of marker proteins can induce the rejection of donor cells through an immune response against marker proteins [14].

In this chapter, we describe the methods for transplantation of primary hepatic progenitor cells after retrorsine treatment and 70% partial hepatectomy in mice [7, 15]. This method can be used for wild-type and nude mice as recipients. As representative data, the results of transplantation of Dlk^+ fetal hepatic progenitor cells derived from heterozygous GFP-Tg mice are presented.

2 Materials

2.1 Preparation of Donor Cells

1. Donor mice (C57BL/6 background, *see* **Note 1**).
2. Sterile phosphate buffered saline (PBS), calcium-free.

3. EGTA buffer (pH 7.5): Hanks' balanced salt solution (without magnesium and calcium) supplemented with 10 mM HEPES and 0.5 mM EGTA [16]. Weigh 8.0 g NaCl, 0.4 g KCl, 0.121 g $Na_2HPO_4{\cdot}12H_2O$, 0.06 g KH_2PO_4, 0.35 g $NaHCO_3$, 1.0 g glucose, 2.383 g HEPES, and 0.19 g EGTA, and transfer to the beaker. Add distilled water to a volume of 900 mL. Mix and adjust pH to 7.5 in a total volume of 1 L with distilled water, and filter through a 0.22 μm bottle-top filter. Store at 4 °C.
4. Collagenase buffer (pH 7.5): Hanks' balanced salt solution (with magnesium sulfate and 5 mM calcium chloride, *see* **Note 2**) supplemented with 10 mM HEPES [16]. Weigh 8.0 g NaCl, 0.4 g KCl, 0.121 g $Na_2HPO_4{\cdot}12H_2O$, 0.06 g KH_2PO_4, 0.35 g $NaHCO_3$, 1.0 g glucose, 2.383 g HEPES, 0.2 g, 0.2 g $MgSO_4{\cdot}7H_2O$, and 0.735 g $CaCl_2{\cdot}2H_2O$, and transfer to the beaker. Add distilled water to a volume of 900 mL. Mix and adjust pH to 7.5 in a total volume of 1 L with distilled water, and filter through a 0.22 μm bottle-top filter. Store at room temperature.
5. Collagenase solution: Dissolve 50 mg collagenase in collagenase buffer in a total volume of 100 mL (final concentration, 0.05%). Filter through a 0.22-μm syringe filter. The solution is dispensed into 15-mL centrifuge tube for a single use, and store at −20 °C. Thaw the solution prior to use.
6. Staining medium: PBS supplemented with 3% fetal calf serum (FCS).
7. Dulbecco's Modified Eagle medium (DMEM) supplemented with 10% FCS.
8. NH_4Cl solution: Dissolve 4.012 g NH_4Cl in distilled water in a total volume of 1 L (final concentration, 75 mM). Filter through a 0.22-μm bottle-top filter.
9. Rat anti-mouse Dlk antibody (*see* **Note 3**).
10. Magnetic cell selection system: Magnet beads conjugated with secondary antibody for rat IgG and appropriate system for each magnetic bead protocol, including positive selection column and magnetic tube holder (*see* **Notes 4** and **5**).
11. Small scissors.
12. Fine small forceps.
13. Nylon mesh with 70-μm pores.

2.2 Recipient Mice and Pretreatment Before Transplantation

1. Mice. C57BL/6 mice or nude mice (male or female) are used in this protocol. Pretreatment of recipient mice starts at 3 or 4 weeks old.

2. 20 mg/mL retrorsine stock solution: Dissolve 40 mg retrorsine in 100% ethanol in a total volume of 2 mL. Store at −20 °C (*see* **Note 6**).
3. Small scissors.
4. Small forceps.
5. Surgical suture.
6. Anesthetic drug (as permitted in your institution).

3 Methods

3.1 Preparation of Donor Cells (Hepatic Progenitor Cells Derived from Fetal Mice)

1. Fetal livers derived from mid-gestational fetal mice (embryonic day 12.5–14.5) are harvested using fine small forceps and placed in sterile PBS. The livers are then transferred to EGTA buffer in a sterile Petri dish and minced using small scissors at room temperature.
2. Transfer the minced livers gently using a pipette into a 15 mL centrifuge tube. Incubate the minced livers in EGTA buffer at 37 °C for 5 min.
3. Centrifuge the collection tube at 150–200 × *g* for 3 min (*see* **Note 7**). Remove the supernatant and tap the centrifugation tube gently to suspend the cell pellet.
4. Suspend with 0.05% collagenase solution (*see* **Note 8**). Incubate the cells at 37 °C for 15 min. After incubation, tap the centrifugation tube vigorously (*see* **Note 9**).
5. Add DMEM with 10% FCS (*see* **Note 10**), and centrifuge the collection tube at 150–200 × *g* for 3 min. Discard the supernatant.
6. Suspend the pellet in 75 mM NH_4Cl solution (*see* **Note 11**), and incubate the cells at 4 °C for 7 min.
7. Tap the collection tube (*see* **Note 12**), and add DMEM with 10% FCS to stop the hemolysis. Centrifuge the collection tube at 150–200 × *g* for 3 min.
8. Discard supernatant and suspend the cells in staining medium. Cell suspensions are filtered using a cell strainer with a 70-μm filter (*see* **Note 13**).
9. Incubate the cells with anti-Dlk antibody at 4 °C for 30 min. After incubation, cells are suspended in staining medium and centrifuged at 150–200 × *g* for 3 min (*see* **Note 4**).
10. Discard supernatant and suspend the cells in staining medium. Incubate the cells with secondary antibody and treat the cells according to the manufacturer's instructions for each magnetic selection system.

11. Suspend the harvested cells in DMEM with 10% FCS. Cell suspensions are filtered using nylon mesh with 70-μm pores.
12. The volume of cell suspension should be adjusted to 0.1–0.15 mL/mouse (*see* **Note 14**). Harvested cells should be transplanted into recipient mice as soon as possible.

3.2 Pretreatment of Recipient Mice

1. To inhibit replication of intrinsic hepatocytes, retrorsine is used in this protocol [7, 15]. Dilute retrorsine stock solution with sterile PBS (*see* **Note 6**) and adjust the final concentration of retrorsine to 4 mg/mL in 20% ethanol/PBS.
2. Intraperitoneally (i.p.) inject 60 mg/kg retrorsine to recipient mice.
3. Treat recipient mice weekly with retrorsine (60 mg/kg i.p.) for 3 weeks (*see* **Note 15**).
4. Three days after the last retrorsine injection, partial hepatectomy and cell transplantation procedures are performed.

3.3 Partial Hepatectomy

1. Perform 70% hepatectomy under general anesthesia using the classic procedure described by Higgins and Anderson [17].
2. Briefly, after the confirmation of deep anesthesia, cut the skin and peritoneum at midline of the abdomen.
3. Ligate the left medial lobe and right medial lobe (*see* **Note 16**) of the liver proximally using surgical silk sutures. Confirm a change of liver color, indicating effective ischemia of the ligated lobes.
4. Remove the left medial lobe and right medial lobe (*see* **Note 17**).
5. Ligate the left lateral lobe of the liver proximally using surgical silk sutures. Confirm a change of liver color, indicating effective ischemia of the ligated lobes.
6. Remove the left lateral lobe.

3.4 Cell Transplantation

1. Loosely ligate the inferior pole of the spleen using surgical silk sutures (*see* **Note 18**).
2. Inject an aliquot of cells suspended in 0.1–0.15 mL of DMEM supplemented with 10% FCS into the splenic pulp from the inferior pole of the spleen using a 30-gauge needle.
3. Tightly ligate the inferior pole of the spleen proximal to the injection point (*see* **Note 19**).
4. Suture the skin and peritoneum using surgical sutures.
5. Sacrifice mice 3–12 weeks after transplantation, at which time donor cells can be clearly detected in recipient livers (Figs. 1 and 2).

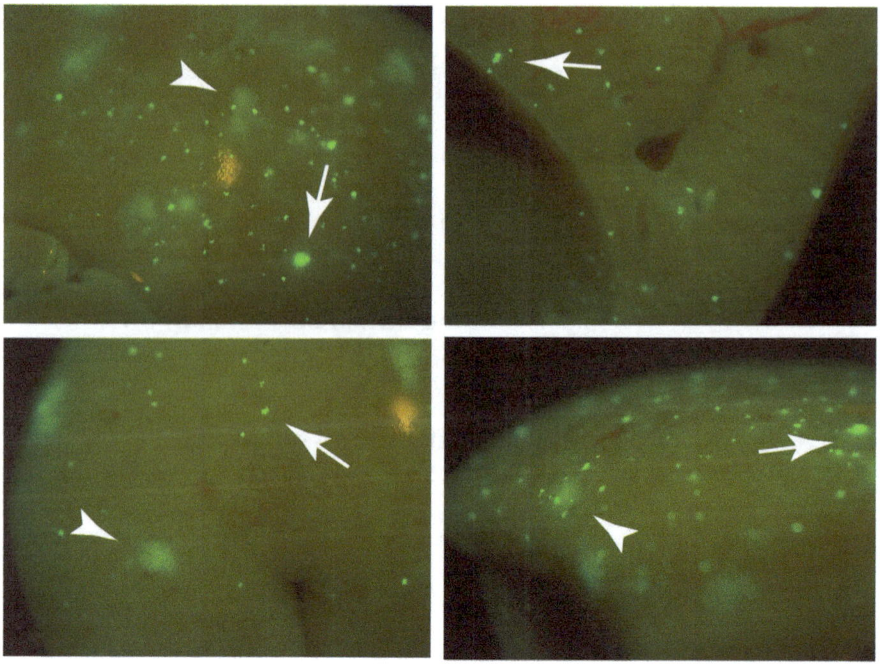

Fig. 1 Transplanted E13.5 Dlk^+ hepatic progenitor cells exhibit liver-repopulating activity in vivo. Dlk^+ cells derived from E13.5 fetal livers were transplanted into nude mice treated with retrorsine before partial hepatectomy. Mice were analyzed 4 weeks after transplantation. Representative fluorescent stereoscopic images of recipient liver are shown. The recipient liver contained many GFP^+ signals. Bright positive signals (arrows) and dull positive signals (arrow heads) were detected, with variation in brightness ascribed to variation in proximity to the liver surface

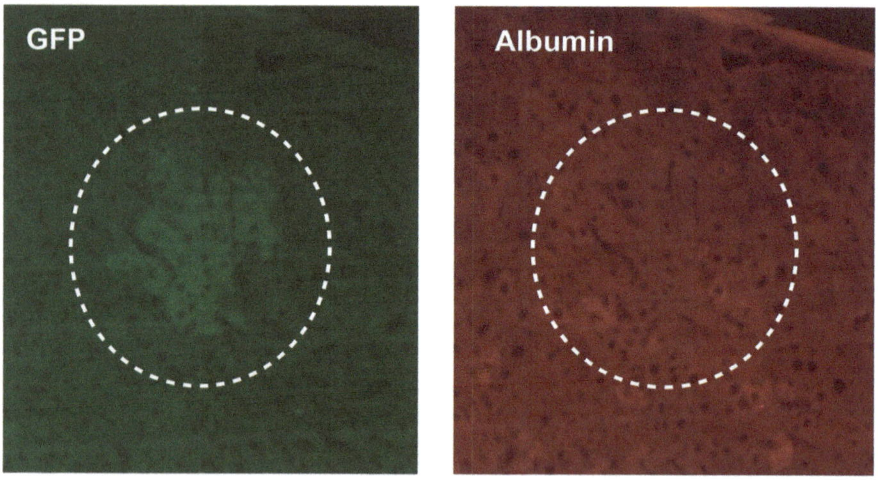

Fig. 2 Microscopic view of GFP^+ donor cells. GFP^+ donor cells were detected as a colony composed of 5–10 cells (dashed circle). Donor-derived cells formed colonies producing GFP (green) and albumin (red), indicating that such cells engrafted as parenchymal cells

4 Notes

1. Donor cells should be tagged with an appropriate marker protein. Heterozygous GFP-Tg mice and heterozygous LacZ/ROSA26-Tg mice have been widely used in previous studies. Donor cells derived from such Tg mice in recipient livers are easily detected using a stereoscopic (fluorescent) microscope or by immunohistological analysis. However, it is possible that overexpression of the marker protein can induce the rejection of donor cells by triggering a recipient immune response against the marker protein. Such immune rejection can be avoided in immunocompromised recipients, such as nude or SCID mice.
2. The final concentration of calcium chloride (5 mM) is higher than that of normal Hanks' balanced salt solution, whereas the final concentration of magnesium sulfate is the same as that of normal Hanks' balanced salt solution.
3. Other appropriate primary antibodies can be used for positive selection of hepatic progenitor cells, such as EpCAM, CD13, and CD133. Negative selection (for hematopoietic cells) can be combined with the positive selection to increase the purity of hepatic progenitor cells.
4. A fluorescence-activated cell sorter (FACS) can be used instead of a magnetic selection system. However, it is possible for FACS-sorted cells to be damaged by sheath pressure and prolonged sorting if the target for the final yield of hepatic progenitors is more than 1×10^5. If target cells are to be purified by FACS, cell sorting should be done after negative selection (for hematopoietic cells) using the magnetic system in order to save time.
5. A centrifuge tube (15 mL) with low adsorption properties is appropriate for collection of hepatic progenitor cells.
6. Dilute 20 mg/mL retrorsine solution fivefold with PBS (−) before injection to recipient mice.
7. To decrease hematopoietic cells in cell suspensions, cells are centrifuged at a relatively low speed compared with blood cell separation.
8. Adjust the volume of collagenase solution depending on the volume of liver cells. When 8–10 livers derived from embryonic day (E) 13.5 mice are harvested, liver cells are suspended with 10 mL collagenase solution.
9. To dissociate the liver cells, tap the tube more vigorously at this step than at previous steps.

10. To inactivate collagenase, serum should be added to the cell suspension.
11. Adjust the volume of NH_4Cl solution depending on the volume of liver cells. When 8–10 livers derived from E13.5 mice are harvested, liver cells are suspended with 7 mL NH_4Cl solution.
12. To induce hemolysis of red blood cells, tap the tube vigorously at this step.
13. To avoid contamination with dead cell cluster, cell suspension should be filtered at this step.
14. The engraftment rate of hepatic progenitor cells in recipient livers is relatively low compared with that of mature hepatocytes. Donor cells are easily detected after transplantation of 1×10^4 adult hepatocytes, whereas it is very difficult to detect the donor cells in the recipient liver after transplantation of less than 1×10^4 hepatic progenitor cells. If the number of harvested cells is less than 1×10^4 cells, in vitro expansion of such cells should be considered. If the number of harvested cells is more than 1×10^5 cells, primary hepatic progenitor cells can be detected using a stereoscopic microscope.
15. Retrorsine treatment should be started when mice are 3–4 weeks old to inhibit proliferation of intrinsic hepatocytes. It is possible that retrorsine treatment is relatively ineffective against older mice.
16. The liver should be ligated gently. The gallbladder is situated between the left medial lobe and right medial lobe. Remove the gallbladder at the same time as liver resection.
17. When ischemia of the ligated lobes is sufficient, there will be little bleeding from the surface of scission.
18. Do not tightly ligate the sutures at this step.
19. To avoid the backflow of the injected cell suspension, tightly ligate with silk sutures (loosely ligated at the previous step) just after injection of cell suspension.

Acknowledgments

This work was supported in part by Grants-in-Aid for Scientific Research from the Ministry of Education, Culture, Sports, Science and Technology in Japan, Japan Agency for Medical Research and Development, and the Japanese Society of Gastroenterology. We thank Prof. H. Nakauchi (the University of Tokyo), Prof. A. Miyajima (the University of Tokyo), and Dr. N. Tanimizu (Sapporo Medical University) for the helpful discussion.

References

1. Rashid T, Kobayashi T, Nakauchi H (2014) Revisiting the flight of Icarus: making human organs from PSCs with large animal chimeras. Cell Stem Cell 15(4):406–409
2. Shiojiri N, Lemire JM, Fausto N (1991) Cell lineages and oval cell progenitors in rat liver development. Cancer Res 51(10):2611–2620
3. Kakinuma S, Nakauchi H, Watanabe M (2009) Hepatic stem/progenitor cells and stem-cell transplantation for the treatment of liver disease. J Gastroenterol 44(3):167–172
4. Tanimizu N, Nishikawa M, Saito H, Tsujimura T, Miyajima A (2003) Isolation of hepatoblasts based on the expression of Dlk/Pref-1. J Cell Sci 116. (Pt 9:1775–1786
5. Nitou M, Sugiyama Y, Ishikawa K, Shiojiri N (2002) Purification of fetal mouse hepatoblasts by magnetic beads coated with monoclonal anti-e-cadherin antibodies and their in vitro culture. Exp Cell Res 279(2):330–343
6. Watanabe T, Nakagawa K, Ohata S, Kitagawa D, Nishitai G, Seo J, Tanemura S, Shimizu N, Kishimoto H, Wada T, Aoki J, Arai H, Iwatsubo T, Mochita M, Watanabe T, Satake M, Ito Y, Matsuyama T, Mak TW, Penninger JM, Nishina H, Katada T (2002) SEKl/MKK4-mediated SAPK/JNK signaling participates in embryonic hepatoblast proliferation via a pathway different from NF-kappaB-induced anti-apoptosis. Dev Biol 250 (2):332–347
7. Kakinuma S, Ohta H, Kamiya A, Yamazaki Y, Oikawa T, Okada K, Nakauchi H (2009) Analyses of cell surface molecules on hepatic stem/progenitor cells in mouse fetal liver. J Hepatol 51(1):127–138
8. Oertel M, Menthena A, Chen YQ, Teisner B, Jensen CH, Shafritz DA (2008) Purification of fetal liver stem/progenitor cells containing all the repopulation potential for normal adult rat liver. Gastroenterology 134(3):823–832
9. Kamiya A, Kakinuma S, Yamazaki Y, Nakauchi H (2009) Enrichment and clonal culture of progenitor cells during mouse postnatal liver development in Mice. Gastroenterology 137 (3):1114–1126
10. Dorrell C, Erker L, Schug J, Kopp JL, Canaday PS, Fox AJ, Smirnova O, Duncan AW, Finegold MJ, Sander M, Kaestner KH, Grompe M (2011) Prospective isolation of a bipotential clonogenic liver progenitor cell in adult mice. Genes Dev 25(11):1193–1203
11. Huch M, Dorrell C, Boj SF, van Es JH, Li VS, van de Wetering M, Sato T, Hamer K, Sasaki N, Finegold MJ, Haft A, Vries RG, Grompe M, Clevers H (2013) In vitro expansion of single Lgr5+ liver stem cells induced by Wnt-driven regeneration. Nature 494(7436):247–250
12. Rhim JA, Sandgren EP, Palmiter RD, Brinster RL (1995) Complete reconstitution of mouse liver with xenogeneic hepatocytes. Proc Natl Acad Sci USA 92(11):4942–4946
13. Overturf K, Al-Dhalimy M, Manning K, Ou CN, Finegold M, Grompe M (1998) Ex vivo hepatic gene therapy of a mouse model of Hereditary Tyrosinemia Type I. Hum Gene Ther 9(3):295–304
14. Maeda H, Shigoka M, Wang Y, Fu Y, Wesson RN, Lin Q, Montgomery RA, Enzan H, Sun Z (2014) Disappearance of GFP-positive hepatocytes transplanted into the liver of syngeneic wild-type rats pretreated with retrorsine. PLoS One 9(5):e95880
15. Dan YY, Riehle KJ, Lazaro C, Teoh N, Haque J, Campbell JS, Fausto N (2006) Isolation of multipotent progenitor cells from human fetal liver capable of differentiating into liver and mesenchymal lineages. Proc Natl Acad Sci U S A 103(26):9912–9917
16. Kost DP, Michalopoulos GK (1991) Effect of 2% dimethyl sulfoxide on the mitogenic properties of epidermal growth factor and hepatocyte growth factor in primary hepatocyte culture. J Cell Physiol 147(2):274–280
17. Higgins G, Anderson R (1931) Experimental pathology of the liver I: restoration of the liver of the white ret following partial surgical removal. Arch Pathol 12:186–202

Chapter 20

Mouse Model for Hepatocellular Carcinoma and Cholangiocarcinoma Originated from Mature Hepatocytes

Masahiro Yamamoto, Bing Xin, and Yuji Nishikawa

Abstract

Liver cancer consists of two main histological subtypes, hepatocellular carcinoma and cholangiocarcinoma, both of which have poor prognosis. Therefore, in searching for new therapeutic targets, adequate mouse models to develop and validate therapeutic strategies are urgently needed. Although there are mouse models of liver cancer, each model has shortcomings. To overcome these shortcomings, a mouse model using a hydrodynamic tail vein injection and the Sleeping Beauty transposon was developed. By inducing stable expression of oncogenes in mouse hepatocytes in vivo, the model can easily induce liver cancer with specific characteristics that depend on the oncogenes used to induce carcinogenesis. Here, we describe the details of the methods to induce hepatocellular carcinoma or cholangiocarcinoma from mouse hepatocytes.

Key words Sleeping Beauty transposon, Hydrodynamic tail vein injection, Hepatocellular carcinoma, Intrahepatic cholangiocarcinoma, Mouse model, Liver tumor, Liver cancer

1 Introduction

Liver cancer is the second most common cause of cancer death worldwide, and its prognosis is very poor (the ratio of mortality to incidence is 0.95) [1]. Liver cancer consists of two main histological subtypes, hepatocellular carcinoma (HCC) and intrahepatic cholangiocarcinoma (CC). The risk factors associated with HCC include hepatitis B virus (HBV) and hepatitis C virus (HCV), alcohol, nonalcoholic fatty liver disease, diabetes, environmental toxin exposure, and obesity [2, 3]. The incidence of CC is highest in Northern Thailand, where infection with liver flukes (*Clonorchis sinensis* and *Opisthorchis viverrini*) leads to chronic biliary injury [4]. In regions without liver fluke infection, the main etiologies of CC are primary sclerosing cholangitis and HBV or HCV infection [4]. Lifestyle-related risk factors, such as smoking, alcohol, and diabetes, also contribute to CC [5].

Naoki Tanimizu (ed.), *Hepatic Stem Cells: Methods and Protocols*, Methods in Molecular Biology, vol. 1905, https://doi.org/10.1007/978-1-4939-8961-4_20,

The therapies for HCC are surgical resection, liver transplantation, ablation, transarterial chemoembolization, and tyrosine kinase inhibitors [6]. Sorafenib, a multiple kinase inhibitor, is used as standard drug therapy for HCC and has some survival benefit [7], but its effects remain limited [8, 9]. Most patients with CC are found in the advanced stage and have poor prognosis despite treatment with current standard chemotherapy [10]. Even for patients who undergo surgical resection, prognosis is still poor due to the high recurrence rate [4]. Therefore, appropriate animal models for liver cancer are necessary to identify therapeutic targets and perform preclinical studies.

Several mouse models of liver cancer are currently available, including chemical carcinogenesis models, genetically engineered mouse models, and xenograft models of human cancer cell lines [11]. The chemical carcinogenesis model is a classical and widely used model, but the genetic abnormalities in these models are usually unknown, and the duration of tumor development is long. Although genetically engineered mouse models are useful models to examine the roles of the modified genes, these models are costly and time-consuming [12]. Furthermore, the important caveat of genetically engineered mouse models is that mutations typically occur in all hepatocytes, while only tumor cells harbor mutations in human cancer. Xenograft models are frequently used to evaluate chemotherapeutic drugs, but the cell lines sometimes do not reflect the major population of the original tumors because of selection during culture [13]. Because xenograft models of human cancer cell lines require immunocompromised mice, the effects of host immunity against tumor cells are neglected in this model.

To overcome these limitations of mouse models, a mouse model with the combination of a hydrodynamic tail vein injection (HTVi) and Sleeping Beauty (SB) transposon-mediated somatic integration of genes has been developed and widely used [12]. This model can induce liver tumors in immunocompetent mice in a short period by known oncogenes. The method can cause long-term expression of oncogenes in mouse hepatocytes and induce liver tumors in vivo.

The standard procedure of HTVi involves the rapid (5–7 s) injection of a large volume (8–10% of body weight) of a plasmid solution via the mouse lateral tail vein [14, 15]. During injection, the solution first accumulates in the right atrium; then, the solution flows retrogradely into the hepatic vein and hepatic sinusoids. The pressure in the sinusoid pushes the plasmid solution into hepatocytes through fenestrated hepatic sinusoidal endothelial cells. By the hydrodynamic mechanism, HTVi can deliver the plasmid specifically to hepatocytes [16–19]. The limitation of the HTVi method for cancer research is that gene expression is only transient, peaking 6 h after transfection and rapidly decreasing to a level less of than 1/1000 in 3 days [14].

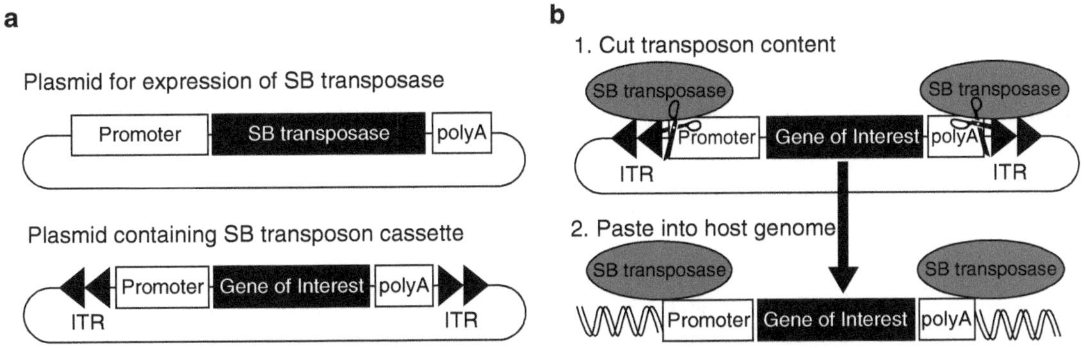

Fig. 1 The Sleeping Beauty (SB) transposon system. (**a**) Components of the Sleeping Beauty transposon system. The SB transposase expression plasmid and a plasmid containing SB transposon cassette. *ITR* inverted terminal repeat. (**b**) Integration of the transposon cassette into a host genome by the SB transposase. SB transposase transfers a gene in the transposon cassette into a host genome in a cut-and-paste manner

Transposons were discovered in the 1940s by Barbara McClintock in the maize genome, and they are known as jumping genes. The Human Genome Project revealed that approximately 45% of the human genome is transposon-derived, but the transposons are inactive [20]. Zoltan Ivics modified a transposon and transposase of the salmonid fish to work in mammalian cells and named the modified transposon the SB transposon [21]. The SB transposon system consists of two components, a transposase-expressing vector and a plasmid with the SB transposon cassette, which contains a promoter, a gene to be expressed, and a poly A sequence that is flanked by inverted terminal repeats (ITRs) (Fig. 1a). When these components are introduced into mouse hepatocytes by the HTVi method, the induced SB transposase recognizes ITRs on the transposon cassette and then cuts the content and pastes it into the mouse hepatocyte genome (Fig. 1b), leading to stable expression of the oncogenes. Integration occurs at any site containing A:T/T:A sequences [22]. With this method, several liver cancer models have been developed that have different phenotypes that are reminiscent of HCC, CC, and hepatoblastoma, depending on the genes used for tumor induction (Table 1).

The cellular origins of HCC and CC are still debated [23, 24]. HCC and CC are thought to be derived from malignant transformation of corresponding parenchymal cells; hepatocytes and intrahepatic bile ducts. Recent evidence has indicated that HCC and CC are derived from transformation of stem/progenitor cells or mature cells subjected to reprogramming, which results in stem/progenitor features [23, 25]. Furthermore, studies on animal models have revealed that CC can originate from mature hepatocytes by transdifferentiation [26–29]. A genome-wide approach in human HCC and CC revealed that CC arising in chronic hepatitis, which is also a risk factor for HCC [30], shows similar gene alteration profiles

Table 1
Sleeping Beauty transposon-mediated liver cancer models

Histology	Genes	Strain	Reference
HCC	myrAKT/HRAS^{G12V}	C57BL/6	[34]
	HRAS^{G12V}/c-Myc	C57BL/6	[34]
	myrAKT/c-Myc	C57BL/6	[34]
	c-MYC/MCL1	FVB/N	[35]
	β-catenin^{S33Y}/KRAS^{G12V} β-catenin^{S45Y}/KRAS^{G12V}	FVB/N	[36]
	β-catenin^{S33Y}/hMet β-catenin^{S45Y}/hMet	FVB/N	[37]
	myrAKT/c-Met	FVB/N	[38]
	myAKT/NRAS^{G12V}	FVB/N	[39]
	myrAKT/Spry2^{Y55F}	FVB/N	[40]
	myrAKT/hMet	FVB/N, C57BL/6	[41]
	myrAKT	FVB/N	[42]
	hMet/Spry2^{Y55F}	FVB/N	[43]
	NRAS^{G12V}/BMI1	FVB/N	[44]
	ΔN90-β-catenin/Spry2^{Y55F}	FVB/N	[45]
	hMET/ΔN90-β-catenin	FVB/N	[46]
Hepatocellular adenoma	myrAKT/ΔN90-β-catenin	FVB/N, C57BL/6	[41]
	hMet/Ccnd1	FVB/N	[47]
CC	myrAKT/YAP^{S127A}	C57BL/6	[28, 48]
	myrAKT/N1ICD	FBV/N	[27]
Biliary cholangioadenoma	myrAKT/N1ICD myrAKT/N2ICD	C57BL/6	[28]
Mixed HCC and CC	PIK3CA^{H1047R}/YAP^{S127A}	FVB/N	[29]
	NRAS^{G12V}	FVB/N(Arf$^{-/-}$)	[49]
Hepatoblastoma	YAP^{S127A}/c-Myc	C57BL/6	[28]
	ΔN90-β-catenin/YAP^{S127A}	FVB/N	[50]
	c-Myc	FVB/N	[51]
Undifferentiated tumor	HRAS^{G12V}/shTrp53	C57BL/6	[52]

HCC hepatocellular carcinoma, *CC* cholangiocarcinoma

to HCC, suggesting that HCC and CC with chronic hepatitis arise from same cellular origin, such as liver progenitor cells [31].

In this chapter, we describe mouse HCC and CC models that originated from mature hepatocytes in vivo by the HTVi and SB transposon system.

2 Materials

2.1 HTVi

1. Plasmids
   (a) pT2-C-Luc/PGK-SB13 (Addgene#20207, Addgene).
   (b) pT3-EF1α-myrAKT-HA (Addgene#31789, Addgene).

(c) pT3-EF1α-FLAG-YAP^{S127A} (Addgene#86497, Addgene).
(d) pT3-EF1α-3xFLAG-N2ICD [28].
(e) Mice: Male C57BL/6 mice, 6–10 weeks old (*see* **Note 1**).

2. 2.5-mL syringe with a Luer Lock.
3. 25G × 1″ (25 mm) or 27G × 1/2″ (13 mm) needle (*see* **Note 2**).
4. Anesthetic: Isoflurane.
5. Inhalational anesthesia apparatus for small animals.
6. 70% ethanol.
7. Gauze.

2.2 Tissue Sampling

1. Phosphate-buffered saline (PBS).
2. Peristaltic pump.
3. Scale to weight mouse (~15–40 g) and mouse liver (~3–15 g).
4. Digital camera.
5. 4% paraformaldehyde/PBS.
6. Liquid nitrogen.
7. Freezer (−80 °C or less).

2.3 Embedding Tissue into Paraffin

1. Tissue processor.
2. Ethanol.
3. Xylene.
4. Paraffin.
5. Embedding center.
6. Tissue mold.
7. Microtome.
8. Slide glass (*see* **Note 3**).

2.4 Hematoxylin and Eosin Stain

1. Mayer's hematoxylin.
2. 1% eosin Y solution (eosin Y diluted to 1% with distilled water).
3. Cover glass (24 × 40 mm).
4. Mounting media; Entellan new (Merck).

2.5 Immunohistochemistry

1. 30% H_2O_2.
2. Methanol.
3. Antigen retrieval solution; Target Retrieval solution, pH 6.0 (DAKO).
4. Electric thermo that can keep water at 98 °C.
5. 0.01% Tween 20 dissolved in PBS (PBST).

6. Blocking reagent; SuperBlock (PBS) Blocking Buffer (Thermo Fisher Scientific).
7. Secondary antibody; DAKO EnVision System.
8. Diaminobenzidine (DAB) solution; ImmPACT™ DAB (Vector Laboratories).

3 Methods

3.1 HTVi

1. Purify the plasmids with the endotoxin-free Maxiprep kit (Qiagen) or equilibrium. Because endotoxin causes adverse or toxic effects on animals, endotoxin must be removed [32]. A high concentration of endotoxin reduces the transfection efficiency in eukaryotic cells [33].
2. For integrating two oncogenic genes, the molar ratio of SB transposase-expressing plasmid and two plasmids carrying oncogenic genes is 1:2:2, and the total amount of plasmid is 25 μg (*see* **Note 4**).
3. Dissolve the plasmids in 2.5 mL of Ringer solution or saline in a sterile manner.
4. Expose the mice to anesthesia with isoflurane.
5. Gather 2.5 mL of the plasmid solution with a 2.5-mL syringe with a Luer Lock.
6. Attach a 25G needle to the syringe and remove any air bubbles from the syringe.
7. Warm the tail of the mouse with hot water (at 40–50 °C) to dilate the lateral tail vein, and clean the insertion site with 70% ethanol.
8. Insert the needle into the lateral tail vein, and rapidly infuse the solution in a continuous manner within 8 s.
9. Press the insertion site with gauze until bleeding stops.
10. Successful injection leads to short-term gene expression in about 30% of hepatocytes.

3.2 Hepatocellular Carcinoma Induced by Activation of the PI3 Kinase and RAS-MAP Kinase Pathways (Fig. 2)

1. Inject a combination of pT3-EF1α-myrAKT-HA and pT3-EF1α-FLAG-HRAS^{G12V} with the SB13-expressing plasmid by the HTVi method.
2. Weigh the mice once per week, and carefully check the mice twice per week to determine whether they have abdominal distension or health problems. Abnormal weight gain or loss indicates the development of liver tumors. Usually, 6–8 weeks after the HTVi, the mice die from liver tumors. However, the timing differs depending on the mouse strain, sex, and maturity of HTVi skill.

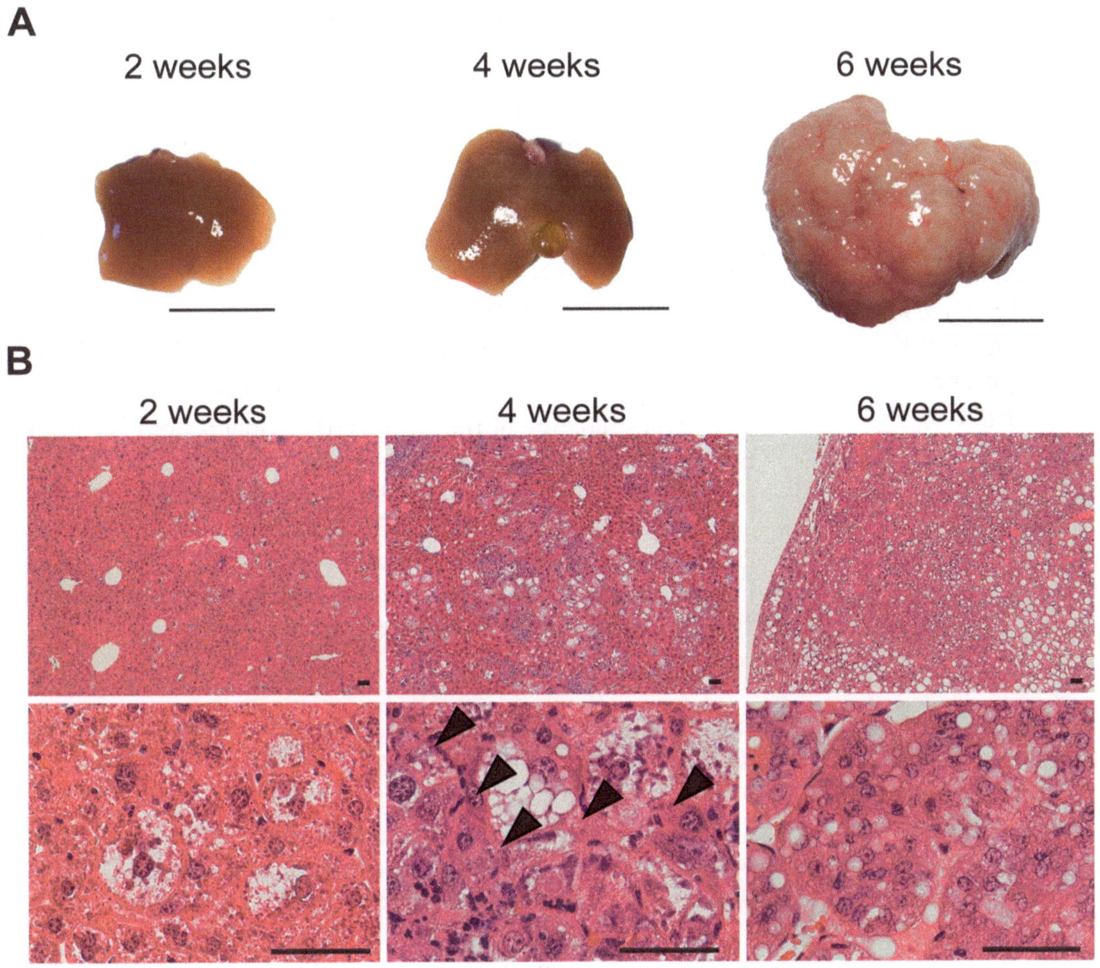

Fig. 2 Hepatocellular carcinoma model induced by myrAKT and $HRAS^{G12V}$. (**a**) Gross appearance of the liver at 2, 4, and 6 weeks after the HTVi. At 6 weeks, multinodular liver tumors develop. Scale bar, 10 mm. (**b**) Hematoxylin and eosin (HE) staining of the liver. At 2 weeks, tumor cells with abundant lipid accumulation in their cytoplasm appear. At 4 weeks, clusters of tumor cells devoid of lipid accumulation (arrowheads) emerge. At 6 weeks, HCCs with scant lipid accumulation develop. The upper panels are enlarged in the lower panels. Scale bar, 50 μm

3. To sample liver tissue, under deep anesthesia, perfuse the liver with PBS via the portal vein by a peristaltic pump and excise the liver. Weigh the whole liver, and calculate the ratio with respect to the body weight. Take macroscopic pictures of the liver with a digital camera, cut the main parts of each lobe of the liver, and fix them with phosphate-buffered 4% paraformaldehyde (4% PFA/PBS) overnight at 4 °C for a paraffin-embedded section. For biochemical analysis, enucleate liver tumors and dice them into small pieces. Then, snap-freeze them using liquid nitrogen. Keep them at −80 °C or less until use.

4. The tumors are macroscopically multinodular. Histologically, the tumors are typical hepatocellular carcinoma. In the early lesion (2 weeks), the tumor cells are lipid-rich; then, in the process of tumor development, the tumor cells lose lipid and proliferate more actively by the action of spontaneously upregulated Myc protein [34].

3.3 Cholangiocarcinoma Induced by Activation of the PI3 Kinase Pathway and Hippo-Yap Pathway (Fig. 3)

1. Inject a combination of pT3-EF1α-myrAKT-HA and pT3-EF1α-FLAG-YAP^{S127A} plasmids with the SB13-expressing vector by the HTVi method.
2. The observation methods are the same as above. Usually, 6–8 weeks after the HTVi, mice die from liver tumors.
3. The earliest change in the AKT/YAP model is lipid accumulation in transformed hepatocytes by activation of the PI3 kinase pathway. At 2 weeks, spindle-shaped cells that are devoid of lipid accumulation proliferate. The cells gradually form ductular structures with an increasing fibrous stroma. The immunohistochemical markers for cholangiocytes, such as cytokeratin 19 (CK19) and grainyhead-like 2 (Grhl2), are expressed in the cells.

3.4 Biliary Cystadenoma Induced by Activation of the PI3 Kinase and Notch Pathways (Fig. 4)

1. Inject a combination of pT3-EF1α-myrAKT-HA and pT3-EF1α-3xFLAG-N2ICD with the SB13-expressing plasmid by the HTVi method.
2. The observation methods are the same as above. Usually, 6–8 weeks after HTVi, multicystic tumors develop. In this model, mice can tolerate severe abdominal distension.
3. Microscopically, the tumors show multicysts lined with cholangiocytic tumor cells that invade less into the surrounding tissue.

3.5 Embedding Tissues in Paraffin Wax and Cutting Thin Sections

1. After incubating tissues in 4% PFA/PBS at 4 °C overnight, transfer the tissues to 70% ethanol and keep the samples at 4 °C until tissue processing.
2. Use a tissue processor to incubate the tissues as follows:
   (a) 70% ethanol, 1 h once.
   (b) 80% ethanol, 1 h once.
   (c) 90% ethanol, 1 h once.
   (d) 95% ethanol, 1 h once.
   (e) 100% ethanol, 1 h three times.
   (f) Xylene, 1 h three times.
   (g) Paraffin wax, 1 h three times at 62 °C.
3. Remove tissues from paraffin wax and transfer them to the embedding center.

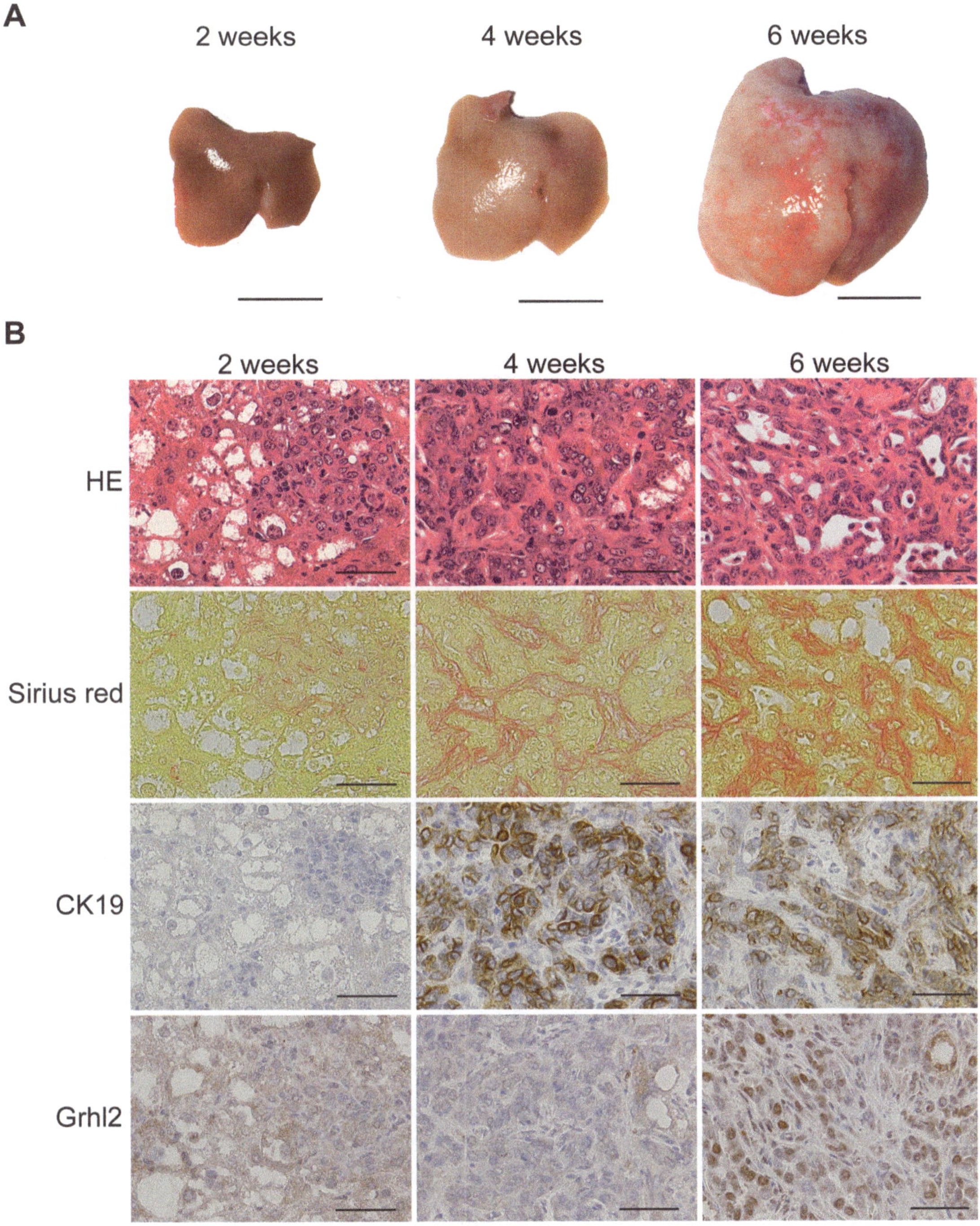

Fig. 3 Cholangiocarcinoma model induced by myrAKT and YAP^{S127A}. (**a**) Gross appearances of the liver at 2, 4, and 6 weeks after the HTVi. Tumors diffusely infiltrate into the liver. Scale bar, 10 mm. (**b**) Microscopic appearances of the AKT/YAP tumors. HE stain, Sirius red stain, and immunohistochemistry for cytokeratin 19 (CK19) and grainyhead-like 2 (Grhl2) at 2, 4, and 6 weeks after the HTVi. Scale bar, 50 μm. Reproduced from Ref. 28 with permission from Elsevier

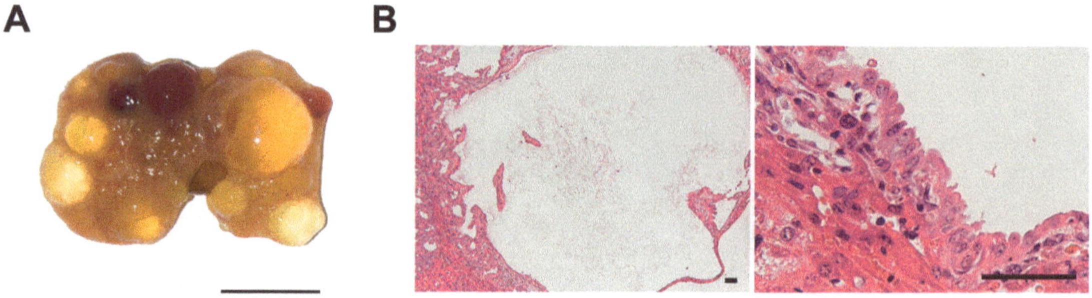

Fig. 4 Biliary cystadenoma model induced by myrAKT and N2ICD. (**a**) Gross appearances of the liver 6 weeks after the HTVi. Scale bar, 10 mm. (**b**) HE staining of the tumors. Scale bar, 50 μm

4. Dispense paraffin wax into an embedding mold and let it sit on a heated surface.
5. Place tissues at an appropriate orientation in the embedding mold, place a cassette on top of the tissue mold, and move the embedding mold to a cold surface of the embedding center to solidify the wax.
6. Use a microtome to cut 3 μm sections from the paraffin block. Place the section into a tissue-sectioning bath at 50 °C for 2–3 s, and mount it onto an appropriate slide glass (*see* **Note 3**).
7. Keep slides at 50 °C overnight to allow sections to adhere to the slides.

3.6 Hematoxylin and Eosin Staining

1. Deparaffinize sections in a deparaffinization series as in the following incubation protocol:
   - (a) 100% xylene, 10 min three times.
   - (b) 50% xylene/50% ethanol, 5 min once.
   - (c) 100% ethanol, 1 min three times.
   - (d) 90% ethanol, 1 min once.
   - (e) 80% ethanol, 1 min once.
   - (f) 70% ethanol, 1 min once.
2. Wash sections with tap water for 3 min.
3. Incubate sections in Mayer's hematoxylin solution for 20 min.
4. Wash sections with running tap water for 10 min.
5. Wash sections with 50% ethanol for 30 s.
6. Incubate sections in eosin solution for 10 min.
7. Dehydrate sections in 70%, 80%, and 90% ethanol for 30 s each, and then move them to 100% ethanol for 2 min three times.
8. To clear, immerse slides in xylene for 5 min four times.
9. Mount the cover glass with mounting media.
10. Dry the slides until the mounting medium solidifies.

Table 2
Primary antibodies for immunohistochemical markers for hepatocellular and cholangiocellular differentiation

Marker for	Antigen	Company	Order no.	Species	Dilution
Hepatocellular differentiation	HNF4αP1	Perseus Proteomics	K9218	Mouse	1:200
	MUP1	Nordic-MUbio	GAM/MUP	Goat	1:500
Cholangiocellular differentiation	CK19	Developed in Dr. Atsushi Miyajima's Laboratory, Tokyo University, Japan		Rabbit	1:4000
	Sox9	Millipore	AB5535	Rabbit	1:20,000
	Grhl2	Sigma	A35809	Rabbit	1:1000

HNF4αP1 hepatocyte nuclear factor α isoform P1, *MUP1* major urinary protein 1, *Sox9* SRY (sex-determining region Y)-box 9

3.7 Immunohistochemistry of Markers for Hepatocytic and Cholangiocytic Differentiation

The antibodies and their optimal dilution for specification of hepatocellular and cholangiocellular differentiation are listed in Table 2. The staining procedure can be performed at room temperature unless otherwise specified.

1. Deparaffinize sections as described in Subheading 3.6.
2. Immerse sections in 3% H_2O_2 dissolved in methanol for 10 min to block endogenous peroxidase.
3. Briefly wash sections with PBST once.
4. Heat sections in a preheated Target Retrieval solution at 98 °C for 40 min in an electric thermos.
5. Remove sections from the electric thermos and cool them to room temperature.
6. Circle the tissues with a hydrophobic pen.
7. Block the tissues with blocking reagent for 30 min in a humidifier chamber. If the source of the primary antibody is mouse, please *see* **Note 5** to avoid secondary antibody to binding to mouse serum.
8. Incubate the tissues with primary antibody dissolved in the blocking reagent for 1 h at room temperature or overnight at 4 °C.
9. Wash the slide with PBST for 5 min twice.
10. Incubate the tissues with secondary antibody for 30 min.
11. Wash the slide with PBST for 5 min twice.
12. Visualize with the DAB solution. During visualization, check the slide under a microscope. When optimal staining is obtained, stop the reaction by putting the slides into tap water.
13. Counterstain slides in Mayer's hematoxylin solution for 1 min.
14. Wash slides with running tap water for 10 min.

15. Rehydrate and clear the slides and then mount a cover glass on them as described in Subheading 3.6.
16. Dry slides until the mounting medium solidifies.

4 Notes

1. Other strains (e.g., FVB, Balb/c, C3H, etc.) can be used. However, susceptibility to tumor development may be different among strains.
2. For smaller mice, a 27G × 1/2″ (13 mm) needle, instead of a 25G needle, can be used.
3. For immunohistochemistry, use slide glasses with a special coat (e.g., Silan) to prevent detachment of tissue sections during the heating step. For usual staining without a heating step in the procedures, glass slides without a special coat are usually sufficient.
4. For a HTVi of three oncogenic genes, the molar ratio of SB transposase-expressing plasmid and three plasmids carrying oncogenic genes is 1:2:2:2, and the total amount of plasmid is 36 μg. However, the ratio and total amounts of plasmid can be changed depending on the purpose of the experiment.
5. For primary antibodies developed in mice, use a staining kit for mouse tissues, such as a Histofine Mouse Stain Kit (Nichirei).

References

1. Ferlay J, Soerjomataram I, Dikshit R, Eser S, Mathers C, Rebelo M, Parkin DM, Forman D, Bray F (2015) Cancer incidence and mortality worldwide: sources, methods and major patterns in GLOBOCAN 2012. Int J Cancer 136(5):E359–E386. https://doi.org/10.1002/ijc.29210
2. Altekruse SF, Henley SJ, Cucinelli JE, McGlynn KA (2014) Changing hepatocellular carcinoma incidence and liver cancer mortality rates in the United States. Am J Gastroenterol 109(4):542–553. https://doi.org/10.1038/ajg.2014.11
3. El-Serag HB, Kanwal F (2014) Epidemiology of hepatocellular carcinoma in the United States: where are we? Where do we go? Hepatology 60(5):1767–1775. https://doi.org/10.1002/hep.27222
4. Razumilava N, Gores GJ (2014) Cholangiocarcinoma. Lancet 383(9935):2168–2179. https://doi.org/10.1016/S0140-6736(13)61903-0
5. Palmer WC, Patel T (2012) Are common factors involved in the pathogenesis of primary liver cancers? A meta-analysis of risk factors for intrahepatic cholangiocarcinoma. J Hepatol 57(1):69–76. https://doi.org/10.1016/j.jhep.2012.02.022
6. Forner A, Reig M, Bruix J (2018) Hepatocellular carcinoma. Lancet 391:1301–1314. https://doi.org/10.1016/s0140-6736(18)30010-2
7. Llovet JM, Ricci S, Mazzaferro V, Hilgard P, Gane E, Blanc JF, de Oliveira AC, Santoro A, Raoul JL, Forner A, Schwartz M, Porta C, Zeuzem S, Bolondi L, Greten TF, Galle PR, Seitz JF, Borbath I, Haussinger D, Giannaris T, Shan M, Moscovici M, Voliotis D, Bruix J, SHARP Investigators Study Group (2008) Sorafenib in advanced hepatocellular carcinoma. N Engl J Med 359(4):378–390. https://doi.org/10.1056/NEJMoa0708857
8. Cheng AL, Kang YK, Chen Z, Tsao CJ, Qin S, Kim JS, Luo R, Feng J, Ye S, Yang TS, Xu J,

Sun Y, Liang H, Liu J, Wang J, Tak WY, Pan H, Burock K, Zou J, Voliotis D, Guan Z (2009) Efficacy and safety of sorafenib in patients in the Asia-Pacific region with advanced hepatocellular carcinoma: a phase III randomised, double-blind, placebo-controlled trial. Lancet Oncol 10(1):25–34. https://doi.org/10.1016/S1470-2045(08)70285-7

9. Inuzuka T, Nishikawa H, Sekikawa A, Takeda H, Henmi S, Sakamoto A, Saito S, Kita R, Kimura T, Osaki Y, Kudo M (2011) Complete response of advanced hepatocellular carcinoma with multiple lung metastases treated with sorafenib: a case report. Oncology 81 (Suppl 1):152–157. https://doi.org/10.1159/000333279

10. Valle J, Wasan H, Palmer DH, Cunningham D, Anthoney A, Maraveyas A, Madhusudan S, Iveson T, Hughes S, Pereira SP, Roughton M, Bridgewater J, ABC Trial Investigators (2010) Cisplatin plus gemcitabine versus gemcitabine for biliary tract cancer. N Engl J Med 362 (14):1273–1281. https://doi.org/10.1056/NEJMoa0908721

11. Heindryckx F, Colle I, Van Vlierberghe H (2009) Experimental mouse models for hepatocellular carcinoma research. Int J Exp Pathol 90(4):367–386. https://doi.org/10.1111/j.1365-2613.2009.00656.x

12. Chen X, Calvisi DF (2014) Hydrodynamic transfection for generation of novel mouse models for liver cancer research. Am J Pathol 184(4):912–923. https://doi.org/10.1016/j.ajpath.2013.12.002

13. Landgraf M, McGovern JA, Friedl P, Hutmacher DW (2018) Rational design of mouse models for cancer research. Trends Biotechnol 36:242–251. https://doi.org/10.1016/j.tibtech.2017.12.001

14. Liu F, Song Y, Liu D (1999) Hydrodynamics-based transfection in animals by systemic administration of plasmid DNA. Gene Ther 6 (7):1258–1266. https://doi.org/10.1038/sj.gt.3300947

15. Zhang G, Budker V, Wolff JA (1999) High levels of foreign gene expression in hepatocytes after tail vein injections of naked plasmid DNA. Hum Gene Ther 10(10):1735–1737. https://doi.org/10.1089/10430349950017734

16. Yokoo T, Kamimura K, Abe H, Kobayashi Y, Kanefuji T, Ogawa K, Goto R, Oda M, Suda T, Terai S (2016) Liver-targeted hydrodynamic gene therapy: Recent advances in the technique. World J Gastroenterol 22 (40):8862–8868. https://doi.org/10.3748/wjg.v22.i40.8862

17. Suda T, Liu D (2007) Hydrodynamic gene delivery: its principles and applications. Mol Ther 15(12):2063–2069. https://doi.org/10.1038/sj.mt.6300314

18. Sawyer GJ, Rela M, Davenport M, Whitehorne M, Zhang X, Fabre JW (2009) Hydrodynamic gene delivery to the liver: theoretical and practical issues for clinical application. Curr Gene Ther 9(2):128–135

19. Bonamassa B, Hai L, Liu D (2011) Hydrodynamic gene delivery and its applications in pharmaceutical research. Pharm Res 28 (4):694–701. https://doi.org/10.1007/s11095-010-0338-9

20. Lander ES, Linton LM, Birren B, Nusbaum C, Zody MC, Baldwin J, Devon K, Dewar K, Doyle M, FitzHugh W, Funke R, Gage D, Harris K, Heaford A, Howland J, Kann L, Lehoczky J, LeVine R, McEwan P, McKernan K, Meldrim J, Mesirov JP, Miranda C, Morris W, Naylor J, Raymond C, Rosetti M, Santos R, Sheridan A, Sougnez C, Stange-Thomann Y, Stojanovic N, Subramanian A, Wyman D, Rogers J, Sulston J, Ainscough R, Beck S, Bentley D, Burton J, Clee C, Carter N, Coulson A, Deadman R, Deloukas P, Dunham A, Dunham I, Durbin R, French L, Grafham D, Gregory S, Hubbard T, Humphray S, Hunt A, Jones M, Lloyd C, McMurray A, Matthews L, Mercer S, Milne S, Mullikin JC, Mungall A, Plumb R, Ross M, Shownkeen R, Sims S, Waterston RH, Wilson RK, Hillier LW, McPherson JD, Marra MA, Mardis ER, Fulton LA, Chinwalla AT, Pepin KH, Gish WR, Chissoe SL, Wendl MC, Delehaunty KD, Miner TL, Delehaunty A, Kramer JB, Cook LL, Fulton RS, Johnson DL, Minx PJ, Clifton SW, Hawkins T, Branscomb E, Predki P, Richardson P, Wenning S, Slezak T, Doggett N, Cheng JF, Olsen A, Lucas S, Elkin C, Uberbacher E, Frazier M, Gibbs RA, Muzny DM, Scherer SE, Bouck JB, Sodergren EJ, Worley KC, Rives CM, Gorrell JH, Metzker ML, Naylor SL, Kucherlapati RS, Nelson DL, Weinstock GM, Sakaki Y, Fujiyama A, Hattori M, Yada T, Toyoda A, Itoh T, Kawagoe C, Watanabe H, Totoki Y, Taylor T, Weissenbach J, Heilig R, Saurin W, Artiguenave F, Brottier P, Bruls T, Pelletier E, Robert C, Wincker P, Smith DR, Doucette-Stamm L, Rubenfield M, Weinstock K, Lee HM, Dubois J, Rosenthal A, Platzer M, Nyakatura G, Taudien S, Rump A, Yang H, Yu J, Wang J, Huang G, Gu J, Hood L, Rowen L, Madan A, Qin S, Davis RW, Federspiel NA, Abola AP, Proctor MJ, Myers RM, Schmutz J, Dickson M, Grimwood J, Cox DR, Olson MV, Kaul R, Raymond C, Shimizu N, Kawasaki K, Minoshima S, Evans GA, Athanasiou M, Schultz R, Roe BA, Chen F,

Pan H, Ramser J, Lehrach H, Reinhardt R, McCombie WR, de la Bastide M, Dedhia N, Blocker H, Hornischer K, Nordsiek G, Agarwala R, Aravind L, Bailey JA, Bateman A, Batzoglou S, Birney E, Bork P, Brown DG, Burge CB, Cerutti L, Chen HC, Church D, Clamp M, Copley RR, Doerks T, Eddy SR, Eichler EE, Furey TS, Galagan J, Gilbert JG, Harmon C, Hayashizaki Y, Haussler D, Hermjakob H, Hokamp K, Jang W, Johnson LS, Jones TA, Kasif S, Kaspryzk A, Kennedy S, Kent WJ, Kitts P, Koonin EV, Korf I, Kulp D, Lancet D, Lowe TM, McLysaght A, Mikkelsen T, Moran JV, Mulder N, Pollara VJ, Ponting CP, Schuler G, Schultz J, Slater G, Smit AF, Stupka E, Szustakowki J, Thierry-Mieg D, Thierry-Mieg J, Wagner L, Wallis J, Wheeler R, Williams A, Wolf YI, Wolfe KH, Yang SP, Yeh RF, Collins F, Guyer MS, Peterson J, Felsenfeld A, Wetterstrand KA, Patrinos A, Morgan MJ, de Jong P, Catanese JJ, Osoegawa K, Shizuya H, Choi S, Chen YJ, Szustakowki J, International Human Genome Sequencing Consortium (2001) Initial sequencing and analysis of the human genome. Nature 409(6822):860–921. https://doi.org/10.1038/35057062

21. Ivics Z, Hackett PB, Plasterk RH, Izsvak Z (1997) Molecular reconstruction of Sleeping Beauty, a Tc1-like transposon from fish, and its transposition in human cells. Cell 91 (4):501–510
22. Izsvak Z, Ivics Z (2004) Sleeping beauty transposition: biology and applications for molecular therapy. Mol Ther 9(2):147–156. https://doi.org/10.1016/j.ymthe.2003.11.009
23. Marquardt JU, Andersen JB, Thorgeirsson SS (2015) Functional and genetic deconstruction of the cellular origin in liver cancer. Nat Rev Cancer 15(11):653–667. https://doi.org/10.1038/nrc4017
24. Sell S (2002) Cellular origin of hepatocellular carcinomas. Semin Cell Dev Biol 13 (6):419–424
25. Oikawa T (2016) Cancer Stem cells and their cellular origins in primary liver and biliary tract cancers. Hepatology 64(2):645–651. https://doi.org/10.1002/hep.28485
26. Sekiya S, Suzuki A (2012) Intrahepatic cholangiocarcinoma can arise from Notch-mediated conversion of hepatocytes. J Clin Invest 122 (11):3914–3918. https://doi.org/10.1172/JCI63065
27. Fan B, Malato Y, Calvisi DF, Naqvi S, Razumilava N, Ribback S, Gores GJ, Dombrowski F, Evert M, Chen X, Willenbring H (2012) Cholangiocarcinomas can originate from hepatocytes in mice. J Clin Invest 122 (8):2911–2915. https://doi.org/10.1172/JCI63212
28. Yamamoto M, Xin B, Watanabe K, Ooshio T, Fujii K, Chen X, Okada Y, Abe H, Taguchi Y, Miyokawa N, Furukawa H, Nishikawa Y (2017) Oncogenic determination of a broad spectrum of phenotypes of hepatocyte-derived mouse liver tumors. Am J Pathol 187:2711–2725. https://doi.org/10.1016/j.ajpath.2017.07.022
29. Li X, Tao J, Cigliano A, Sini M, Calderaro J, Azoulay D, Wang C, Liu Y, Jiang L, Evert K, Demartis MI, Ribback S, Utpatel K, Dombrowski F, Evert M, Calvisi DF, Chen X (2015) Co-activation of PIK3CA and Yap promotes development of hepatocellular and cholangiocellular tumors in mouse and human liver. Oncotarget 6(12):10102–10115. https://doi.org/10.18632/oncotarget.3546
30. Lee CH, Chang CJ, Lin YJ, Yeh CN, Chen MF, Hsieh SY (2009) Viral hepatitis-associated intrahepatic cholangiocarcinoma shares common disease processes with hepatocellular carcinoma. Br J Cancer 100(11):1765–1770. https://doi.org/10.1038/sj.bjc.6605063
31. Fujimoto A, Furuta M, Totoki Y, Tsunoda T, Kato M, Shiraishi Y, Tanaka H, Taniguchi H, Kawakami Y, Ueno M, Gotoh K, Ariizumi S, Wardell CP, Hayami S, Nakamura T, Aikata H, Arihiro K, Boroevich KA, Abe T, Nakano K, Maejima K, Sasaki-Oku A, Ohsawa A, Shibuya T, Nakamura H, Hama N, Hosoda F, Arai Y, Ohashi S, Urushidate T, Nagae G, Yamamoto S, Ueda H, Tatsuno K, Ojima H, Hiraoka N, Okusaka T, Kubo M, Marubashi S, Yamada T, Hirano S, Yamamoto M, Ohdan H, Shimada K, Ishikawa O, Yamaue H, Chayama K, Miyano S, Aburatani H, Shibata T, Nakagawa H (2016) Whole-genome mutational landscape and characterization of noncoding and structural mutations in liver cancer. Nat Genet 48(5):500–509. https://doi.org/10.1038/ng.3547
32. Kim MJ, Ahituv N (2013) The hydrodynamic tail vein assay as a tool for the study of liver promoters and enhancers. Methods Mol Biol 1015:279–289. https://doi.org/10.1007/978-1-62703-435-7_18
33. Butash KA, Natarajan P, Young A, Fox DK (2000) Reexamination of the effect of endotoxin on cell proliferation and transfection efficiency. BioTechniques 29(3):610–614, 616, 618–619
34. Xin B, Yamamoto M, Fujii K, Ooshio T, Chen X, Okada Y, Watanabe K, Miyokawa N, Furukawa H, Nishikawa Y (2017) Critical role of Myc activation in mouse hepatocarcinogenesis induced by the activation of AKT and RAS

pathways. Oncogene 36(36):5087–5097. https://doi.org/10.1038/onc.2017.114

35. Mendez-Lucas A, Li X, Hu J, Che L, Song X, Jia J, Wang J, Xie C, Driscoll PC, Tschaharganeh DF, Calvisi DF, Yuneva M, Chen X (2017) Glucose catabolism in liver tumors induced by c-MYC can be sustained by various PKM1/PKM2 ratios and pyruvate kinase activities. Cancer Res 77(16):4355–4364. https://doi.org/10.1158/0008-5472.CAN-17-0498
36. Tao J, Zhang R, Singh S, Poddar M, Xu E, Oertel M, Chen X, Ganesh S, Abrams M, Monga SP (2017) Targeting beta-catenin in hepatocellular cancers induced by coexpression of mutant beta-catenin and K-Ras in mice. Hepatology 65(5):1581–1599. https://doi.org/10.1002/hep.28975
37. Tao J, Xu E, Zhao Y, Singh S, Li X, Couchy G, Chen X, Zucman-Rossi J, Chikina M, Monga SP (2016) Modeling a human hepatocellular carcinoma subset in mice through coexpression of met and point-mutant beta-catenin. Hepatology 64(5):1587–1605. https://doi.org/10.1002/hep.28601
38. Hu J, Che L, Li L, Pilo MG, Cigliano A, Ribback S, Li X, Latte G, Mela M, Evert M, Dombrowski F, Zheng G, Chen X, Calvisi DF (2016) Co-activation of AKT and c-Met triggers rapid hepatocellular carcinoma development via the mTORC1/FASN pathway in mice. Sci Rep 6:20484. https://doi.org/10.1038/srep20484
39. Ho C, Wang C, Mattu S, Destefanis G, Ladu S, Delogu S, Armbruster J, Fan L, Lee SA, Jiang L, Dombrowski F, Evert M, Chen X, Calvisi DF (2012) AKT (v-akt murine thymoma viral oncogene homolog 1) and N-Ras (neuroblastoma ras viral oncogene homolog) coactivation in the mouse liver promotes rapid carcinogenesis by way of mTOR (mammalian target of rapamycin complex 1), FOXM1 (forkhead box M1)/SKP2, and c-Myc pathways. Hepatology 55(3):833–845. https://doi.org/10.1002/hep.24736
40. Wang C, Delogu S, Ho C, Lee SA, Gui B, Jiang L, Ladu S, Cigliano A, Dombrowski F, Evert M, Calvisi DF, Chen X (2012) Inactivation of Spry2 accelerates AKT-driven hepatocarcinogenesis via activation of MAPK and PKM2 pathways. J Hepatol 57(3):577–583. https://doi.org/10.1016/j.jhep.2012.04.026
41. Stauffer JK, Scarzello AJ, Andersen JB, De Kluyver RL, Back TC, Weiss JM, Thorgeirsson SS, Wiltrout RH (2011) Coactivation of AKT and beta-catenin in mice rapidly induces formation of lipogenic liver tumors. Cancer Res 71(7):2718–2727. https://doi.org/10.1158/0008-5472.CAN-10-2705
42. Calvisi DF, Wang C, Ho C, Ladu S, Lee SA, Mattu S, Destefanis G, Delogu S, Zimmermann A, Ericsson J, Brozzetti S, Staniscia T, Chen X, Dombrowski F, Evert M (2011) Increased lipogenesis, induced by AKT-mTORC1-RPS6 signaling, promotes development of human hepatocellular carcinoma. Gastroenterology 140(3):1071–1083. https://doi.org/10.1053/j.gastro.2010.12.006
43. Lee SA, Ladu S, Evert M, Dombrowski F, De Murtas V, Chen X, Calvisi DF (2010) Synergistic role of Sprouty2 inactivation and c-Met up-regulation in mouse and human hepatocarcinogenesis. Hepatology 52(2):506–517. https://doi.org/10.1002/hep.23681
44. Xu CR, Lee S, Ho C, Bommi P, Huang SA, Cheung ST, Dimri GP, Chen X (2009) Bmi1 functions as an oncogene independent of Ink4A/Arf repression in hepatic carcinogenesis. Mol Cancer Res 7(12):1937–1945. https://doi.org/10.1158/1541-7786.MCR-09-0333
45. Lee SA, Ho C, Roy R, Kosinski C, Patil MA, Tward AD, Fridlyand J, Chen X (2008) Integration of genomic analysis and in vivo transfection to identify sprouty 2 as a candidate tumor suppressor in liver cancer. Hepatology 47(4):1200–1210. https://doi.org/10.1002/hep.22169
46. Tward AD, Jones KD, Yant S, Cheung ST, Fan ST, Chen X, Kay MA, Wang R, Bishop JM (2007) Distinct pathways of genomic progression to benign and malignant tumors of the liver. Proc Natl Acad Sci U S A 104(37):14771–14776. https://doi.org/10.1073/pnas.0706578104
47. Patil MA, Lee SA, Macias E, Lam ET, Xu C, Jones KD, Ho C, Rodriguez-Puebla M, Chen X (2009) Role of cyclin D1 as a mediator of c-Met- and beta-catenin-induced hepatocarcinogenesis. Cancer Res 69(1):253–261. https://doi.org/10.1158/0008-5472.CAN-08-2514
48. Zhang S, Song X, Cao D, Xu Z, Fan B, Che L, Hu J, Chen B, Dong M, Pilo MG, Cigliano A, Evert K, Ribback S, Dombrowski F, Pascale RM, Cossu A, Vidili G, Porcu A, Simile MM, Pes GM, Giannelli G, Gordan J, Wei L, Evert M, Cong W, Calvisi DF, Chen X (2017) Pan-mTOR inhibitor MLN0128 is effective against intrahepatic cholangiocarcinoma in mice. J Hepatol 67(6):1194–1203. https://doi.org/10.1016/j.jhep.2017.07.006
49. Carlson CM, Frandsen JL, Kirchhof N, McIvor RS, Largaespada DA (2005) Somatic integration of an oncogene-harboring Sleeping Beauty transposon models liver tumor

development in the mouse. Proc Natl Acad Sci U S A 102(47):17059–17064. https://doi.org/10.1073/pnas.0502974102

50. Tao J, Calvisi DF, Ranganathan S, Cigliano A, Zhou L, Singh S, Jiang L, Fan B, Terracciano L, Armeanu-Ebinger S, Ribback S, Dombrowski F, Evert M, Chen X, Monga SPS (2014) Activation of beta-catenin and Yap1 in human hepatoblastoma and induction of hepatocarcinogenesis in mice. Gastroenterology 147(3):690–701. https://doi.org/10.1053/j.gastro.2014.05.004

51. Chow EK, Fan LL, Chen X, Bishop JM (2012) Oncogene-specific formation of chemoresistant murine hepatic cancer stem cells. Hepatology 56(4):1331–1341. https://doi.org/10.1002/hep.25776

52. Ju HL, Ahn SH, Kim DY, Baek S, Chung SI, Seong J, Han KH, Ro SW (2013) Investigation of oncogenic cooperation in simple liver-specific transgenic mouse models using noninvasive in vivo imaging. PLoS One 8(3):e59869. https://doi.org/10.1371/journal.pone.0059869

Chapter 21

Mouse Model for Cholangiocarcinoma from Peribiliary Glands

Hayato Nakagawa, Nobumi Suzuki, and Kazuhiko Koike

Abstract

A good mouse model is mandatory for elucidating carcinogenic mechanisms and identifying the cellular origin of cancer. Although the lack of an appropriate mouse model has hampered investigation of extrahepatic cholangiocarcinoma (ECC), we recently established a novel mouse model of biliary injury-related ECC by ductal cell-specific activation of Kras and deletion of transforming growth factor (TGF) β receptor type 2 and E-cadherin. Using this mouse model, we identified that peribiliary glands, which are considered a biliary epithelial stem cell niche, are potential cellular origins of ECC. Furthermore, we established an extrahepatic biliary organoid-derived xenograft cholangiocarcinoma (CC) model by lentiviral induction of Cre in organoids. This organoid system recreated the in vivo conditions and facilitated analysis of carcinogenesis. In this chapter, we describe the protocol used to establish our mouse model of ECC derived from peribiliary glands and our extrahepatic biliary organoid-derived xenograft model of CC.

Key words Extrahepatic cholangiocarcinoma, Biliary organoid, Peribiliary gland, Inflammation, E-cadherin

1 Introduction

Cholangiocarcinoma (CC) is a highly malignant cancer, and its incidence is increasing worldwide [1]. CC is commonly classified into intrahepatic and extrahepatic CC (ICC and ECC, respectively) according to anatomical location, and ECC is further divided into hilar and distal CC [2]. As 10–20% of CC cases occur in the presence of chronic biliary injury and inflammation, such as primary sclerosing cholangitis, liver flukes, and hepatitis viral infection, chronic inflammation plays a key role in CC development [1]. Importantly, recent studies using next-generation sequencing technology have provided comprehensive mutation profiling of CC, identifying an altered mutation spectrum depending on anatomical subtype and underlying etiology [3, 4].

Recent genetic lineage-tracing studies have provided new insights into the cellular origin of CC. Although CC has been

Naoki Tanimizu (ed.), *Hepatic Stem Cells: Methods and Protocols*, Methods in Molecular Biology, vol. 1905,
https://doi.org/10.1007/978-1-4939-8961-4_21, © Springer Science+Business Media, LLC, part of Springer Nature 2019

considered to originate from biliary epithelial cells (BECs), mature hepatocytes also give rise to ICC via activation of Notch signaling [5, 6]. In regard to ECC, the anatomical and immunohistochemical analyses revealed that peribiliary glands (PBGs), clusters of epithelial cells residing in the submucosal compartment of extrahepatic bile ducts (EHBDs), form epithelial networks within the walls of EHBDs and are a potential stem cell niche of BECs; thus, PBGs have attracted attention as a potential origin of ECC [7, 8].

A good mouse model is necessary to identify the cellular origin of cancer. Several mouse models of ICC have been established and have contributed to our understanding of the molecular carcinogenic mechanisms of ICC. However, because most of these models were generated by liver-specific genetic manipulation or hepatotoxins, CC development was limited in the intrahepatic and perihilar areas [5, 9–11]. The lack of an appropriate mouse model has been a considerable barrier to the study of ECC.

Recently, we established a new mouse model of ECC by ductal cell-specific gene manipulation using *K19*^{CreERT} mice, in which tamoxifen-inducible Cre ERT was knocked into the endogenous K19 locus [12]. Based on recent human studies showing frequent genetic alterations of Ras and transforming growth factor (TGF)β/SMAD signaling pathways in ICC and ECC, mice exhibiting ductal cell-specific Kras activation and TGFβ receptor type 2 (TGFβR2) inactivation were generated by crossing *LSL-Kras*^{G12D}, *Tgfbr2*$^{flox/flox}$, and *K19*^{CreERT} mice (*KT-K19*^{CreERT}). *KT-K19*^{CreERT} mice showed only mild hyperplasia of BECs in the extrahepatic bile duct; however, additional loss of E-cadherin by crossing *KT-K19*^{CreERT} mice with *CDH1*$^{flox/flox}$ mice (*KTC-K19*^{CreERT}) resulted in development of periductal infiltrating-type ECC, the progression of which depended on biliary epithelial injury-induced inflammatory and regenerative responses. Time-course analysis of EHBDs after tamoxifen administration revealed that recombined BECs lining the bile duct lumen had died and detached as a result of E-cadherin loss, whereas recombined cells survived in the PBGs, which provide a cancer stem cell niche. Cell lineage tracing suggested that PBGs are the cellular origin of ECC. Furthermore, we established an extrahepatic biliary organoid-derived xenograft CC model by lentiviral induction of Cre into organoids derived from the EHBD of Cre-negative *KTC* mice. Interestingly, detachment of BECs was also seen in Cre^{+} *KTC* organoids, which was similar to that seen in the in vivo model. Therefore, this EHBD organoid system recreated the in vivo conditions and facilitated analysis of carcinogenesis induced by genetic manipulations in BECs separately from the tumor microenvironment. Using these novel mouse and organoid models, we showed that dying BECs release interleukin-33, which stimulated inflammation and subsequent regeneration of the EHBD by PBGs via type 2 innate lymphoid cells, eventually leading to ECC [12]. Thus, these models are useful tools for investigating carcinogenic mechanisms, cellular origin, and potential therapeutic targets of ECC.

Here, we describe the protocols used to establish our mouse model of ECC derived from PBGs and our extrahepatic biliary organoid-derived xenograft CC model.

2 Materials

2.1 In Vivo Model of ECC

1. *KTC-K19*^{CreERT} mice: Cross *K19*^{CreERT}, *LSL-Kras*$^{G12D+/-}$, *Tgfbr2*$^{flox/flox}$, and *CDH1*$^{flox/flox}$ mice [13–15]. All mice were of the C57BL/6 genetic background.
2. Tamoxifen: 20 mg/mL in corn oil containing 10% ethanol. Mix for 30 min using a tube rotor.

2.2 Extrahepatic Biliary Organoid-Derived Xenograft CC Model

1. Cre-negative *KTC* mice: Crossing *LSL-Kras*$^{G12D+/-}$, *Tgfbr2*$^{flox/flox}$, and *CDH1*$^{flox/flox}$ mice.
2. Nude mice.
3. Phosphate-buffered saline (PBS)
4. Digestion buffer: 4% fetal calf serum, 1 mg/mL collagenase D, 0.5 g/mL dispase, 40 μg/mL DNase I, and 1 mM EDTA in RPMI.
5. Organoid culture medium: Advanced Dulbecco's Modified Eagle Medium/F-12 supplemented with B27 (1:50) and N2 (1:100), 1 mM *N*-acetylcysteine, 10 mM nicotinamide, 50 ng/mL epidermal growth factor, 1 μg/mL Rspo1, 100 ng/mL Noggin, 100 ng/mL fibroblast growth factor 10, 10% Wnt3a-conditioned medium, and penicillin/streptomycin.
6. The ROCK inhibitor Y27632 (final 10 μM).
7. Matrigel.
8. Cell Recovery Solution (BD Biosciences).
9. 29-gauge insulin syringe.
10. Puromycin (final 5 μg/mL).

3 Methods

3.1 In Vivo Model of ECC

1. Tamoxifen (200 mg/kg) was administered by oral gavage to *KTC-K19*^{CreERT} mice (7–8 weeks old) for 3 consecutive days. This protocol can induce gene recombination (~40%) in all parts of the biliary tree, including the EHBD, PBGs, GB, perihilar bile duct (BD), and intrahepatic bile duct.
2. On day 7, surface BECs started to detach from the BD wall.

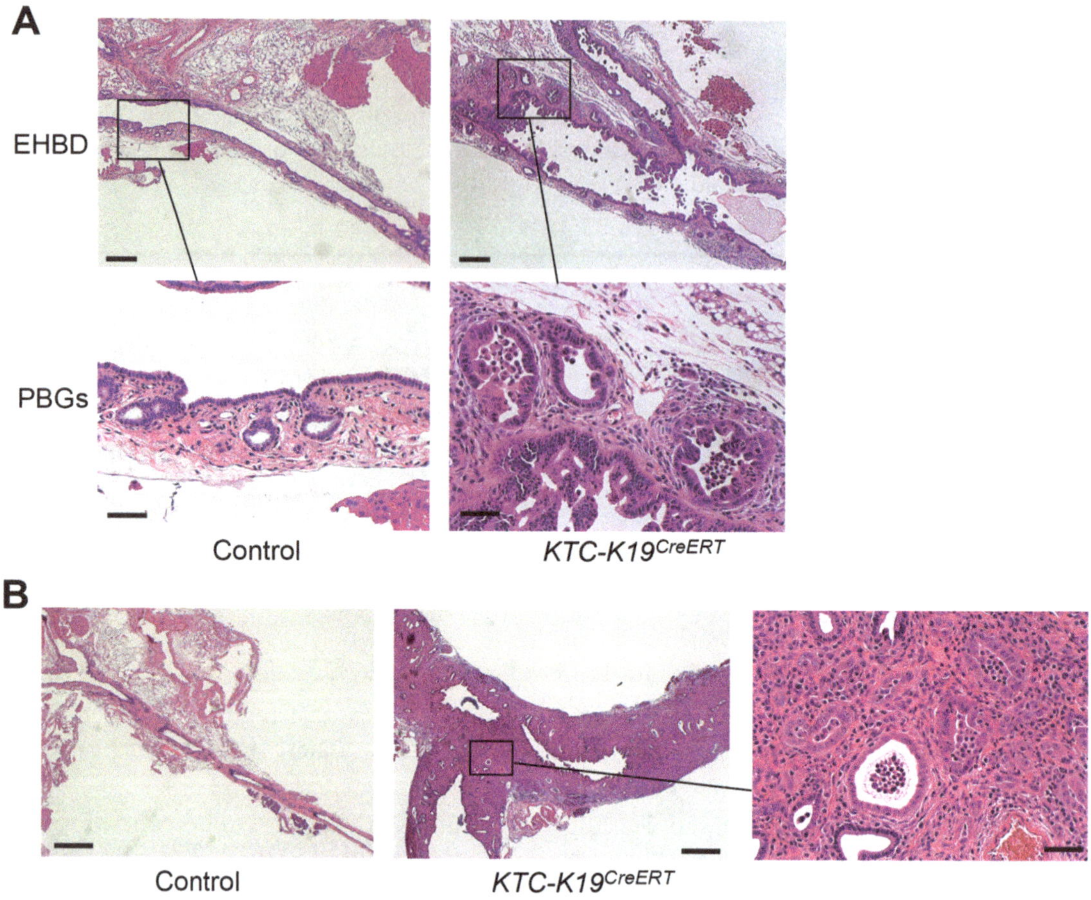

Fig. 1 Histological findings of extrahepatic bile duct (EHBD) in *KTC-K19*^{CreERT} mice. (**a**) Hematoxylin and eosin (H&E)-stained images of EHBDs and PBGs from Cre-negative control and *KTC-K19*^{CreERT} mice at 10 days after tamoxifen administration (scale bar: upper panels, 200 μm; lower panels, 50 μm). Right panel shows enlarged and dysplastic peribiliary glands (PBGs). (**b**) H&E-stained images of EHBDs from Cre-negative control and *KTC-K19*^{CreERT} mice at 3.5 weeks after tamoxifen administration (scale bar: left two panels, 500 μm; right panel, 50 μm)

3. On day 10, many shedding BECs were observed in the BD lumen, and inflammatory cells had infiltrated the subepithelial connective tissue. The PBGs were enlarged and morphologically dysplastic, accompanied by mitotic figures (Fig. 1a).
4. On day 15, although the shedding of surface BECs had subsided, the cancerous glands had spread through the subepithelial area, accompanied by a fibrotic response.
5. On day 20, most of the histological features of periductal infiltrating-type ECC were apparent (Fig. 1b) (*see* **Note 1**).
6. All *KTC-K19*^{CreERT} mice died within 4 weeks of tamoxifen administration mainly from liver failure or respiratory failure caused by lung adenocarcinoma.

3.2 Extrahepatic Biliary Organoid-Derived Xenograft CC Model

3.2.1 Culture of EHBD Organoids

1. Isolate the EHBD tissue from Cre-negative *KTC* mice and place tissue in a 10-cm petri dish filled with PBS.
2. Remove the adhered interstitial and pancreatic tissue from the EHBD using forceps with gentle touching.
3. Cut the EHBD tissue down the middle and divide it into two parts.
4. Place the EHBD tissue in a 1.5-mL tube containing 1 mL digestion buffer.
5. Incubate for 30 min at 37 °C.
6. Gently disperse the EHBD tissue by pipetting.
7. Centrifuge at 300 × *g* for 5 min at 4 °C. Discard the supernatant.
8. Place the 1.5-mL tube containing EHBD cells on ice and add 1 mL cold PBS. Vortex for 10 s.
9. Centrifuge at 300 × *g* for 5 min at 4 °C. Discard the supernatant.
10. Place the 1.5-mL tube containing EHBD cells on ice. Add 80 μL ice-cold Matrigel and mix by pipetting using a cold 200-μL tip with a large-end diameter (cut off end of the tip) (*see* **Note 2**). Do not generate bubbles while mixing.
11. Seed EHBD cells in a 40 μL drop in the center of each well of a 24-well plate.
12. Carefully place the 24-well plate in a 37 °C incubator for 10 min.
13. Add 500 μL culture medium supplemented with ROCK inhibitor to each well.
14. Organoid formation was observed 1–2 days after plating. EHBD organoids formed a ball-like structure and were composed of a single layer of the biliary epithelium (Fig. 2a, b).
15. Refresh the culture medium every 2–3 days and passage the EHBD organoids every 5–10 days.
16. When passaging organoids, remove the culture medium and add 1 mL Cell Recovery Solution to dissolve the Matrigel.
17. Place the solution in a 1.5-mL tube and incubate 10 min on ice to precipitate the organoids by gravity. Remove the upper portion of the supernatant (~700 μL), which contains dead cells and debris.
18. Add 700 μL Cell Recovery Solution and suspend the organoids well.
19. Pass the solution through a 29-gauge insulin syringe once or twice.

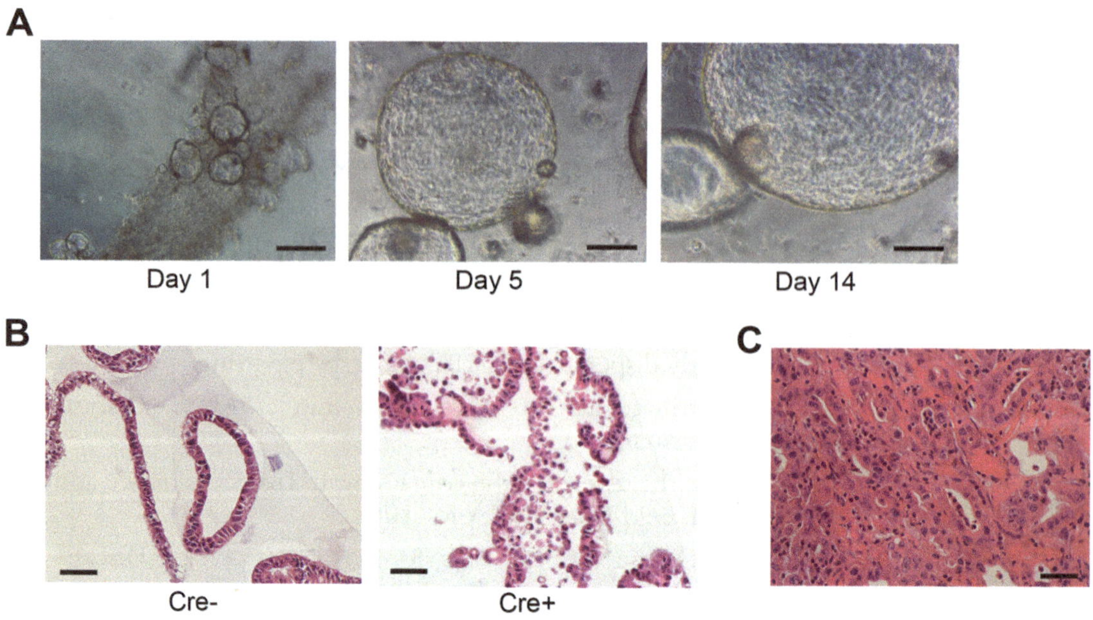

Fig. 2 Extrahepatic biliary organoid-derived xenograft CC model. (**a**) Micrograph of biliary organoids cultured from EHBDs of mice (scale bar: 250 μm). (**b**) H&E-stained images of EHBD organoids from Cre-negative *KTC* mice infected with Cre-expressing or control lentivirus (scale bar: 50 μm). (**c**) H&E-stained image of an organoid-derived tumor (scale bar: 50 μm)

20. Centrifuge at 600 × *g* for 5 min at 4 °C. Discard the supernatant.
21. Add 400 μL cold PBS and suspend well.
22. Transfer approximately 800 organoid fragments to a new 1.5-mL tube on ice.
23. Centrifuge at 600 × *g* for 5 min at 4 °C. Discard the supernatant.
24. Add 160 μL Matrigel and suspend well.
25. Seed 40 μL organoid solution into each well of a 24-well plate.
26. Carefully place the 24-well plate in a 37 °C incubator for 10 min.
27. Add 500 μL culture medium supplemented with ROCK inhibitor into each well.
28. Place the plate in a 37 °C incubator.

3.2.2 Induction of Gene Recombination in EHBD Organoids Using a Cre-Expressing Lentivirus

1. Resolve Matrigel and wash organoids in the same way with Subheading 3.2.1, **steps 16–23**.
2. Add 15 μL lentivirus solution (titer ≥1 × 10^7 transduction units/mL) and suspend well.
3. Incubate 30 min on ice.

4. Add 160 μL Matrigel and suspend well.
5. Seed 40 μL organoid solution into each well of a 24-well plate.
6. Carefully place the 24-well plate in a 37 °C incubator for 10 min.
7. Add 500 μL culture medium supplemented with ROCK inhibitor into each well.
8. Place the plate in a 37 °C incubator.
9. After 2–3 days, replace the culture medium with 500 μL medium containing the selection drug (e.g., 5 μg/mL puromycin).
10. After 2 weeks of selection, confirm gene recombination by polymerase chain reaction or Western blot. Organoids with successful recombination of *LSL-Kras*^{G12D}, *Tgfbr2*$^{flox/flox}$, and *CDH*$^{flox/flox}$ alleles will exhibit shedding BECs (Fig. 2b).

3.2.3 Extrahepatic Biliary Organoid-Derived Xenograft CC Model

1. Culture EHBD organoids from Cre-negative *KTC* mice and induce gene recombination as described above.
2. Seed 100 organoid fragments in each well of a 24-well plate.
3. After 2 days, remove the medium.
4. Add 1 mL cold PBS to each well and mix with the Matrigel containing organoids.
5. Place the solution in a 1.5-mL tube on ice.
6. Centrifuge at 600 × *g* for 5 min at 4 °C.
7. Discard the supernatant and leave approximately 100 μL Matrigel/PBS mixture containing organoids.
8. Add 100 μL Matrigel/PBS mixture (1:1) and suspend well. At this stage, 200 ul suspension remains.
9. Inject 100 μL of the organoid suspension subcutaneously in two sites at the back of nude mice using an ice-cold 29-gauge insulin syringe.
10. After 2 months, approximately two-thirds of the organoids will have formed tumors with a similar histology to that of ECC in *KTC-K19*^{CreERT} mice (Fig. 2c).

4 Notes

1. ECC is induced more stably in male mice (100%) than in female mice (80%).
2. Matrigel should be handled on ice as much as possible to prevent it from gelling.

Acknowledgments

This study was supported by the Japan Society for the Promotion of Science KAKENHI grant number 15K09039, Astellas Foundation for Research on Metabolic Disorders, Nakayama Cancer Research Institute, Okinaka Memorial Institute for Medical Research, Research Grant of the Princess Takamatsu Cancer Research Fund, and Foundation for Promotion of Cancer Research in Japan.

References

1. Rizvi S, Gores GJ (2013) Pathogenesis, diagnosis, and management of cholangiocarcinoma. Gastroenterology 145(6):1215–1229. https://doi.org/10.1053/j.gastro.2013.10.013
2. Malhi H, Gores GJ (2006) Cholangiocarcinoma: modern advances in understanding a deadly old disease. J Hepatol 45(6):856–867. https://doi.org/10.1016/j.jhep.2006.09.001
3. Chan-On W, Nairismagi ML, Ong CK, Lim WK, Dima S, Pairojkul C, Lim KH, McPherson JR, Cutcutache I, Heng HL, Ooi L, Chung A, Chow P, Cheow PC, Lee SY, Choo SP, Tan IB, Duda D, Nastase A, Myint SS, Wong BH, Gan A, Rajasegaran V, Ng CC, Nagarajan S, Jusakul A, Zhang S, Vohra P, Yu W, Huang D, Sithithaworn P, Yongvanit P, Wongkham S, Khuntikeo N, Bhudhisawasdi V, Popescu I, Rozen SG, Tan P, Teh BT (2013) Exome sequencing identifies distinct mutational patterns in liver fluke-related and non-infection-related bile duct cancers. Nat Genet 45(12):1474–1478. https://doi.org/10.1038/ng.2806
4. Nakamura H, Arai Y, Totoki Y, Shirota T, Elzawahry A, Kato M, Hama N, Hosoda F, Urushidate T, Ohashi S, Hiraoka N, Ojima H, Shimada K, Okusaka T, Kosuge T, Miyagawa S, Shibata T (2015) Genomic spectra of biliary tract cancer. Nat Genet 47(9):1003–1010. https://doi.org/10.1038/ng.3375
5. Sekiya S, Suzuki A (2012) Intrahepatic cholangiocarcinoma can arise from Notch-mediated conversion of hepatocytes. J Clin Invest 122(11):3914–3918. https://doi.org/10.1172/JCI63065
6. Fan B, Malato Y, Calvisi DF, Naqvi S, Razumilava N, Ribback S, Gores GJ, Dombrowski F, Evert M, Chen X, Willenbring H (2012) Cholangiocarcinomas can originate from hepatocytes in mice. J Clin Invest 122(8):2911–2915. https://doi.org/10.1172/JCI63212
7. Dipaola F, Shivakumar P, Pfister J, Walters S, Sabla G, Bezerra JA (2013) Identification of intramural epithelial networks linked to peribiliary glands that express progenitor cell markers and proliferate after injury in mice. Hepatology 58(4):1486–1496. https://doi.org/10.1002/hep.26485
8. Lanzoni G, Cardinale V, Carpino G (2016) The hepatic, biliary, and pancreatic network of stem/progenitor cell niches in humans: a new reference frame for disease and regeneration. Hepatology 64(1):277–286. https://doi.org/10.1002/hep.28326
9. Xu X, Kobayashi S, Qiao W, Li C, Xiao C, Radaeva S, Stiles B, Wang RH, Ohara N, Yoshino T, LeRoith D, Torbenson MS, Gores GJ, Wu H, Gao B, Deng CX (2006) Induction of intrahepatic cholangiocellular carcinoma by liver-specific disruption of Smad4 and Pten in mice. J Clin Invest 116(7):1843–1852. https://doi.org/10.1172/JCI27282
10. Ikenoue T, Terakado Y, Nakagawa H, Hikiba Y, Fujii T, Matsubara D, Noguchi R, Zhu C, Yamamoto K, Kudo Y, Asaoka Y, Yamaguchi K, Ijichi H, Tateishi K, Fukushima N, Maeda S, Koike K, Furukawa Y (2016) A novel mouse model of intrahepatic cholangiocarcinoma induced by liver-specific Kras activation and Pten deletion. Sci Rep 6:23899. https://doi.org/10.1038/srep23899
11. Guest RV, Boulter L, Kendall TJ, Minnis-Lyons SE, Walker R, Wigmore SJ, Sansom OJ, Forbes SJ (2014) Cell lineage tracing reveals a biliary origin of intrahepatic cholangiocarcinoma. Cancer Res 74(4):1005–1010. https://doi.org/10.1158/0008-5472.CAN-13-1911
12. Nakagawa H, Suzuki N, Hirata Y, Hikiba Y, Hayakawa Y, Kinoshita H, Ihara S, Uchino K, Nishikawa Y, Ijichi H, Otsuka M, Arita J, Sakamoto Y, Hasegawa K, Kokudo N, Tateishi K, Koike K (2017) Biliary epithelial injury-induced regenerative response by IL-33

promotes cholangiocarcinogenesis from peribiliary glands. Proc Natl Acad Sci U S A 114 (19):E3806–E3815. https://doi.org/10.1073/pnas.1619416114

13. Means AL, Xu Y, Zhao A, Ray KC, Gu G (2008) A CK19(CreERT) knockin mouse line allows for conditional DNA recombination in epithelial cells in multiple endodermal organs. Genesis 46(6):318–323. https://doi.org/10.1002/dvg.20397
14. Ijichi H, Chytil A, Gorska AE, Aakre ME, Fujitani Y, Fujitani S, Wright CV, Moses HL (2006) Aggressive pancreatic ductal adenocarcinoma in mice caused by pancreas-specific blockade of transforming growth factor-beta signaling in cooperation with active Kras expression. Genes Dev 20(22):3147–3160. https://doi.org/10.1101/gad.1475506
15. Nakagawa H, Hikiba Y, Hirata Y, Font-Burgada J, Sakamoto K, Hayakawa Y, Taniguchi K, Umemura A, Kinoshita H, Sakitani K, Nishikawa Y, Hirano K, Ikenoue T, Ijichi H, Dhar D, Shibata W, Akanuma M, Koike K, Karin M, Maeda S (2014) Loss of liver E-cadherin induces sclerosing cholangitis and promotes carcinogenesis. Proc Natl Acad Sci U S A 111 (3):1090–1095. https://doi.org/10.1073/pnas.1322731111

INDEX

Naoki Tanimizu (ed.), *Hepatic Stem Cells: Methods and Protocols*, Methods in Molecular Biology, vol. 1905, https://doi.org/10.1007/978-1-4939-8961-4,

Kolibri 360 GmbH
Ferdinand-Jühlke-Straße 7
99095 Erfurt, Deutschland
produktsicherheit@kolibri360.de